FREE

ALSO BY CHRIS ANDERSON

The Long Tail:
Why the Future of Business Is Selling Less of More

FREE

THE FUTURE OF
A RADICAL PRICE

CHRIS ANDERSON

HYPERION

New York

Library of Congress Cataloging-in-Publication Data

Anderson, Chris
   Free : the future of a radical price / Chris Anderson.
      p.   cm.
   ISBN 978-1-4013-2290-8
   1. Marketing.   2. Success in business   I. Title.
   HF5415.A6197 2009
   658.8'16—dc22

                                                              2009012333

Hyperion books are available for special promotions and premiums. For details contact the HarperCollins Special Markets Department in the New York office at 212-207-7528, fax 212-207-7222, or email spsales@harpercollins.com.

FIRST EDITION

10 9 8 7 6 5 4 3 2 1

THIS LABEL APPLIES TO TEXT STOCK

We try to produce the most beautiful books possible, and we are also extremely concerned about the impact of our manufacturing process on the forests of the world and the environment as a whole. Accordingly, we've made sure that all of the paper we use has been certified as coming from forests that are managed to ensure the protection of the people and wildlife dependent upon them.

To Anne

CONTENTS

FREECONOMICS AND THE FREE WORLD

LIST OF SIDEBARS

FREE

PROLOGUE

IN NOVEMBER 2008, the surviving members of the original Monty Python team, stunned by the extent of digital piracy of their videos, issued a very stern announcement on YouTube:

> For 3 years you YouTubers have been ripping us off, taking tens of thousands of our videos and putting them on YouTube. Now the tables are turned. It's time for us to take matters into our own hands.
>
> We know who you are, we know where you live and we could come after you in ways too horrible to tell. But being the extraordinarily nice chaps we are, we've figured a better way to get our own back: We've launched our own Monty Python channel on YouTube.
>
> No more of those crap quality videos you've been posting. We're giving you the real thing—high quality videos delivered straight from our vault. What's more, we're taking our most viewed clips and uploading brand new high quality versions. And what's even more, we're letting you see absolutely everything for free. So there!
>
> But we want something in return.

None of your driveling, mindless comments. Instead, we want you to click on the links, buy our movies & TV shows and soften our pain and disgust at being ripped off all these years.

Three months later, the results of this rash experiment with Free were in. Monty Python's DVDs had climbed to No. 2 on Amazon's Movies and TV best-sellers list, with increased sales of 23,000 percent. So there!

Free worked, and worked brilliantly. More than 2 million people watched the clips on YouTube as word of mouth spread and parents introduced their children to the Black Knight and the Dead Parrot Sketch. Thousands of viewers were reminded how much they loved Monty Python and wanted more, so they ordered the DVDs. Response videos, mashups, and remixes spread, and a new generation learned the proper meaning of "Killer Rabbit." And all this cost Monty Python essentially nothing, since YouTube paid all the bandwidth and storage costs, such as they were.

What's surprising about this example is how unsurprising it is. There are countless other cases just like this online, where pretty much everything is given away for free in some version with the hopes of selling something else—or, even more frequently, with no expectation of pay at all.

I'm typing these words on a $250 "netbook" computer, which is the fastest growing new category of laptop. The operating system happens to be a version of free Linux, although it doesn't matter since I don't run any programs but the free Firefox Web browser. I'm not using Microsoft Word, but rather free Google Docs, which has the advantage of making my drafts available to me wherever I am, and I don't have to worry about backing them up since Google takes care of that for me. Everything else I do on this computer is free, from my email to my Twitter feeds. Even the wireless access is free, thanks to the coffee shop I'm sitting in.

And yet Google is one of the most profitable companies in America, the "Linux ecosystem" is a $30 billion industry, and the coffee shop seems to be selling $3 lattes as fast as they can make them.

Therein lies the paradox of Free: People are making lots of money charging nothing. Not nothing for everything, but nothing for enough that we have essentially created an economy as big as a good-sized country around the price of $0.00. How did this happen and where is it going?

That's the central question of this book.

For me, it started with a loose end in *The Long Tail*. My first book was about the new shape of consumer demand, when everything is available and we can choose from the infinite aisle rather than just the best-seller bin. The abundant marketplace of the Long Tail was enabled by the unlimited "shelf space" of the Internet, which is the first distribution system in history that is as well suited for the niche as for the mass, for the obscure as well as the mainstream. The result was the birth of a wildly diverse new culture and a threat to the institutions of the existing one, from mainstream media to music labels.

There's only one way you can have unlimited shelf space: if that shelf space costs nothing. The near-zero "marginal costs" of digital distribution (that is, the additional cost of sending out another copy beyond the "fixed costs" of the required hardware with which to do it) allow us to be indiscriminate in what we use it for—no gatekeepers are required to decide if something deserves global reach or not. And out of that free-for-all came the miracle of today's Web, the greatest accumulation of human knowledge, experience, and expression the world has ever seen. So that's what free shelf space can do. As I marveled over the consequences, I started thinking more about Free, and realized just how far it had spread. It didn't just explain the explosion of variety online, it defined the pricing there, too. What's more, this "free" wasn't just a marketing gimmick like the free samples and prizes inside that we're used to in traditional retail. This free seemed to have no strings attached: It wasn't just a lure for a future sale, but genuinely gratis. Most of us depend on one or more Google services every day, but they never show up on our credit card. No meter ticks as you use Facebook. Wikipedia costs you nothing.

Twenty-first-century Free is different from twentieth-century Free.

Somewhere in the transition from atoms to bits, a phenomenon that we thought we understood was transformed. "Free" became Free.

Surely economics must have something to say about this, I thought. But I couldn't find anything. No theories of gratis, or pricing models that went to zero. (In fairness, some do exist, as later research would reveal. But they were mostly obscure academic discussions of "two-sided markets" and, as we'll see in the economics chapter, nearly forgotten theories from the nineteenth century.) Somehow an economy had emerged around Free before the economic model that could describe it.

Thus this book, an exploration of a concept that is in the midst of radical evolution. As I came to learn, Free is both a familiar concept and a deeply mysterious one. It is as powerful as it is misunderstood. The Free that emerged over the past decade is different from the Free that came before, but how and why are rarely explored. What's more, today's Free is full of apparent contradictions: You *can* make money giving things away. There really *is* a free lunch. Sometimes you get *more* than you pay for.

This was a fun book to write. It took me from the patent medicine makers of late-nineteenth-century America to the pirate markets of China. I dived into the psychology of gifts and the morality of waste. I started a project on the side to try out new business models around electronics where the intellectual property is free (a model known as open source hardware). I got to brainstorm with my publishers on the many ways to make this book itself free in most of its forms, while still creating ways for everyone who helped produce it to get paid.

In some ways, this was a public research project, as *The Long Tail* had been. I previewed the thesis in an article in *Wired* and blogged about it as I had with *The Long Tail*. But it took a different path, more in my own head than in a collective conversation with contributors online. This book is more driven by history and narrative, and it is as much about Free's past as it is about its future. My research took me as often to archives and eighteenth-century psychology texts as it did to the latest Web phenomena. And so I found myself in more of a traditional writer's mode, of solitary studying and typing with earphones on in Starbucks, as God intended.

When I wasn't writing, I was traveling, talking to people about Free. I found that the idea that you could create a huge global economy around a base price of zero was invariably polarizing, but the one common factor was that nearly everyone had their doubts. At risk of ageist generalization, there were broadly two camps of skeptics: those over thirty and those below. The older critics, who had grown up with twentieth-century Free, were rightly suspicious: Surely "free" is nothing of the sort—we all pay sooner or later. Not only is it not new, but it's the oldest marketing gimmick in the book. When you hear "free," reach for your wallet.

The younger critics had a different response: "Duh!" This is the Google Generation, and they've grown up online simply assuming that everything digital is free. They have internalized the subtle market dynamics of near-zero marginal cost economics in the same way that we internalize Newtonian mechanics when we learn to catch a ball. The fact that we are now creating a global economy around the price of zero seemed too self-evident to even note.

With that, I realized that this was a perfect subject for a book. Any topic that can divide critics equally into two opposite camps—"totally wrong" and "so obvious"—has got to be a good one. I hope that those who read this book, even if they start in one of those camps, will end in neither. Free is not new, but it *is* changing. And it is doing so in ways that are forcing us to rethink some of our basic understandings of human behavior and economic incentives.

Those who understand the new Free will command tomorrow's markets and disrupt today's—indeed, they're already doing it. This book is about them and what they're teaching us. It is about the past and future of a radical price.

1

THE BIRTH OF FREE

THERE'S NO GETTING AROUND IT: Gelatin comes from flesh and bones. It's the translucent, glutinous substance that skims to the top when you boil meat. But if you collect enough of it and purify it, adding color and flavor, it becomes something else: Jell-O. A clean powder in a packet, far removed from its abattoir origins of marrow and connective tissue.

We don't think much about the origins of Jell-O today, but in the late 1800s, if you wanted to put a jiggly treat on your dinner table, you had to make it the hard way: putting off-cuts in a stewpot and waiting a half day for the hydrolyzed collagen to emerge from the gristle.

In 1895, Pearle Wait, a carpenter in LeRoy, New York, with a side business of patent medicine packaging, sat at his kitchen table poking at a bowl of gelatin. He'd been wanting to get into the then-new packaged foods business and thought this might be the stuff, if only he could figure out how to make it more appealing. Although glue-makers had been producing it for decades as a by-product of their animal rendering, it had yet to prove popular with American consumers. For good reason: It was a lot of work for a pretty small reward.

Wait wondered if there might be a way to take gelatin more mainstream. Earlier efforts to sell prepackaged powdered gelatin, including by the inventor of the process, Peter Cooper (of Cooper Union fame), sold it plain and unflavored on the argument that this was the most flexible form; cooks could add their own flavors. But Wait thought that preflavored gelatins might sell better, so he mixed in fruit juices, along with sugar and food dyes. The jelly took on the color and flavor of the fruits—orange, lemon, raspberry, and strawberry—creating something that looked, smelled, and tasted appealing. Colorful, light, and delightful to play with, it was a treat that could add jiggly, translucent fun to almost any meal. To distance the stuff further from its abattoir origins, his wife, May, renamed it Jell-O. They boxed it up to sell.

But it didn't sell. Jell-O was too foreign a food and too unknown a brand for turn-of-the-century consumers. Kitchen traditions were still based on Victorian recipes, where every food type had its place. Was this new jelly a salad ingredient or a dessert?

For two years, Wait kept trying to stir up interest in Jell-O, with little success. Eventually, in 1899, he gave up and sold the trademark—name, hyphen, and all—to Orator Frank Woodward, a fellow townsman. The price was $450.

Woodward was a natural salesman, and he had settled in the right place. LeRoy had become something of a nineteenth-century huckster hotbed, best known for its patent medicine makers. Woodward sold plenty of miracle cures and was creative with plaster of paris, too. He marketed plaster target balls for marksmen and invented a plaster laying nest for chickens that was infused with an anti-lice powder.

But even Woodward's firm, the Genesee Pure Food Company, struggled to find a market for powdered gelatin. It was a new product category with an unknown brand name in an era where general stores sold almost all products from behind the counter and customers had to ask for them by name. The Jell-O was manufactured in a nearby factory run by Andrew Samuel Nico. Sales were so slow and disheartening for the new product that on one gloomy day, while contemplating a huge stack of unsold Jell-O boxes, Woodward offered Nico the whole business for $35. Nico refused.

The main problem was that consumers didn't understand the product or what they could do with it. And without consumer demand, merchants wouldn't stock it. Manufacturers of other products in the new packaged ingredient business, such as Arm & Hammer baking soda and Fleischmann's yeast, often bundled recipe books with their boxes. Woodward figured a usage guidebook might help create demand for Jell-O, too, but how to get them out there? Nobody was buying the boxes in the first place.

So in 1902 Woodward and his marketing chief, William E. Humelbaugh, tried something new. First, they crafted a three-inch ad to run in *Ladies' Home Journal,* at a cost of $336. Rather optimistically proclaiming Jell-O "America's Most Famous Dessert," the ad explained the appeal of the product: This new dessert "could be served with the simple addition of whipped cream or thin custard. If, however, you desire something very fancy, there are hundreds of delightful combinations that can be quickly prepared."

Then, to illustrate all those richly varied combinations, Genesee printed up tens of thousands of pamphlets with Jell-O recipes and gave them to its salesmen to distribute to homemakers for free.

This cleverly got around the salesmen's chief problem. As they traveled around the country in their buggies, they were prohibited from selling door-to-door in most towns without a costly traveling salesman's license. But the cookbooks were different—giving things away wasn't selling. They could knock on doors and just hand the woman of the house a free recipe book, no strings attached. Printing paper was cheap compared to making Jell-O. They couldn't afford to give out free samples of the product itself, so they did the next best thing: free information that could only be used if the consumer bought the product.

After blanketing a town with the booklets, the salesmen would then go to the local merchants and advise them that they were about to get a wave of consumers asking for a new product called Jell-O, which they would be wise to stock. The boxes of Jell-O in the back of the buggies finally started to move.

By 1904, the campaign had turned into a runaway success. Two years later Jell-O hit a million dollars in annual sales. The company introduced the "Jell-O Girl" in its ads, and the pamphlets grew into Jell-O "bestseller" recipe books. In some years Genesee printed as many as 15 million of the free books, and in the company's first twenty-five years it printed and distributed an estimated quarter billion free cookbooks door-to-door, across the country. Noted artists such as Norman Rockwell, Linn Ball, and Angus MacDonald contributed colored illustrations to the cookbooks. Jell-O had become a fixture in the American kitchen and a household name.

Thus was born one of the most powerful marketing tools of the twentieth century: giving away one thing to create demand for another. What Woodward understood was that "free" is a word with an extraordinary ability to reset consumer psychology, create new markets, break old ones, and make almost any product more attractive. He also figured out that "free" didn't mean profitless. It just meant that the route from product to revenue was indirect, something that would become enshrined in the retail playbook as the concept of a "loss leader."

KING GILLETTE

At the same time, the most famous example of this new marketing method was in the works a few hundred miles north, in Boston. At the age of forty, King Gillette was a frustrated inventor, a bitter anticapitalist, and a salesman of cork-lined bottle caps. Despite ideas, energy, and wealthy parents, he had little to show for his work. He blamed the evils of market competition. Indeed, in 1894 he had published a book, *The Human Drift*, which argued that all industry should be taken over by a single corporation owned by the public and that millions of Americans should live in a giant city called Metropolis powered by Niagara Falls. His boss at the bottle cap company, meanwhile, had just one piece of advice: Invent something people use and throw away.

One day, while he was shaving with a straight razor that was so worn it could no longer be sharpened, the idea came to him. What if the blade could be made of a thin metal strip? Rather than spending time maintaining the blades, men could simply discard them when they became dull. A few years of metallurgy experimentation later, the disposable-blade safety razor was born.

But it didn't take off immediately. In its first year, 1903, Gillette sold a total of 51 razors and 168 blades. Over the next two decades, he tried every marketing gimmick he could think of. He put his own face on the package, making him both legendary and, some people believed, fictional. He sold millions of razors to the army at a steep discount, hoping the habits soldiers developed at war would carry over to peacetime. He sold razors in bulk to banks so they could give them away with new deposits ("shave and save" campaigns). Razors were bundled with everything from Wrigley's gum to packets of coffee, tea, spices, and marshmallows.

The freebies helped to sell those products, but the tactic helped Gillette even more. By selling cheaply to partners who would give away the razors, which were useless by themselves, he was creating demand for disposable blades. It was just like Jell-O (whose cookbooks were the "razors" to the gelatin "blades"), but even more tightly linked. Once hooked on disposable razor blades, you were a daily customer for life.

Interestingly, the idea that Gillette, the company, gave away the razors is mostly urban myth. The only recorded examples were with the introduction of the Trak II in the 1970s, when the company gave away a cheap version of the razor with a nonreplaceable blade. Its more usual model was to sell razors at a low margin to partners, such as banks, who would typically give them away as part of promotions. Gillette made its real profit from the high margin on the blades.

A few billion blades later, this business model is now the foundation of entire industries: Give away the cell phone, sell the monthly plan; make the video game console cheap and sell expensive games; install fancy coffeemakers in offices at no charge so you can sell managers expensive coffee sachets.

Starting from these experiments at the beginning of the twentieth century, Free fueled a consumer revolution that defined the next hundred years. The rise of Madison Avenue and the arrival of the supermarket made consumer psychology a science and Free the tool of choice. "Free-to-air" radio and television (the term used for signals sent over the airways that anyone can receive without charge) united a nation and created the mass market. Free was the rallying cry of the modern marketer, and the consumer never failed to respond.

TWENTY-FIRST-CENTURY FREE

Now, at the beginning of the twenty-first century, we're inventing a new form of Free, and this one will define the next era just as profoundly. The new form of Free is not a gimmick, a trick to shift money from one pocket to another. Instead, it's driven by an extraordinary new ability to lower the costs of goods and services close to zero. While the last century's Free was a powerful marketing method, this century's Free is an entirely new economic model.

This new form of Free is based on the economics of bits, not atoms. It is a unique quality of the digital age that once something becomes software, it inevitably becomes free—in cost, certainly, and often in price. (Imagine if the price of steel had dropped so close to zero that King Gillette could give away both razor and blade, and make his money on something else entirely—shaving cream?) And it's creating a multibillion-dollar economy—the first in history—where the primary price is zero.

In the atoms economy, which is to say most of the stuff around us, things tend to get more expensive over time. But in the bits economy, which is the online world, things get cheaper. The atoms economy is inflationary, while the bits economy is deflationary.

The twentieth century was primarily an atoms economy. The twenty-first century will be equally a bits economy. Anything free in the atoms economy must be paid for by something else, which is why so much traditional free feels like bait and switch—it's you paying, one way or another. But free in the bits economy can be *really* free, with money often taken out of the equation altogether. People are rightly suspicious of Free in the atoms economy, and rightly trusting of Free in the bits economy.

Intuitively, they understand the difference between the two economies, and why Free works so well online.

A decade and a half into the great online experiment, free has become the default, and pay walls the route to obscurity. In 2007, the *New York Times* went free online, as did much of the *Wall Street Journal*, using a clever hybrid model that made stories free to those who wanted to share them online, in blog posts or other social media. Musicians from Radiohead to Nine Inch Nails now routinely give away their music online, realizing that Free lets them reach more people and create more fans, some of whom attend their concerts and even—gasp—pay for premium versions of the music. The fastest-growing parts of the gaming industry are ad-supported casual games online and free-to-play massively multiplayer online games.

The rise of "freeconomics" is being driven by the underlying technologies of the digital age. Just as Moore's Law dictates that a unit of computer processing power halves in price every two years, the price of bandwidth and storage is dropping even faster. What the Internet does is combine all three, compounding the price declines with a triple play of technology: processors, bandwidth, and storage. As a result, the net annual deflation rate of the online world is nearly 50 percent, which is to say that whatever it costs YouTube to stream a video today will cost half as much in a year. The trend lines that determine the cost of doing business online all point the same way: to zero. No wonder the prices online all go the same way.

George Gilder, whose 1990 book, *Microcosm,* was the first to explore the economics of bits, puts this in historical context:

> In every industrial revolution, some key factor of production is drastically reduced in cost. Relative to the previous cost to achieve that function, the new factor is virtually free. [Thanks to steam,] physical force in the Industrial Revolution became virtually free compared to getting it from animal muscle power or human muscle power. Suddenly you could do things you could not afford to do before. You could make a factory work 24 hours a day

churning out products in a way that was just incomprehensible before.

Today the most interesting business models are in finding ways to make money around Free. Sooner or later every company is going to have to figure out how to use Free or compete with Free, one way or another. This book is about how to do that.

First, we'll look at the history of Free and why it has such power over our choices. Then we'll see how digital economics has revolutionized Free, turning it from a marketing gimmick into an economic force, including the new business models it enables. Finally, we'll dive into the underlying principles of freeconomics: how it works, where it works, and why it's so often misunderstood and feared. But to start, what does "free" really mean?

2

FREE 101

A Short Course on a Most Misunderstood Word

"FREE" CAN MEAN MANY THINGS, and that meaning has changed over the years. It raises suspicions, yet has the power to grab attention like almost nothing else. It is almost never as simple as it seems, yet it is the most natural transaction of all. If we are now building an economy around Free, we should start by understanding what it is and how it works.

Let's begin with the definition. In Latinate languages, such as French, Spanish, and Italian, "free" is less convoluted because it is not a single word. Instead, it is two words, one derived from the Latin *liber* ("freedom") and the other from the Latin *gratis* (contraction of *gratiis*, "for thanks," hence, "without recompense," or zero price). In Spanish, for instance, *libre* is a good thing (freedom of speech, etc.) while *gratis* is often suspected of being a marketing gimmick.

In English, though, the two words are mushed together into a single word. This has marketing advantages: the positive "freedom" connotation lowers our defenses to sales tricks. But it also introduces ambiguity. (Which is why English speakers sometimes use "gratis" for emphasis, to underscore that something is *really* free.)

In the open source software world, which is both free (encouraging use and reuse) *and* free (no charge), people distinguish between the two like this: "Free as in beer vs. free as in speech." (Inevitably, some over-clever types thought it would be funny to reambiguate this by releasing a beer recipe under a share-and-share-alike license and then charging for the finished product at software conferences. Geeks!)

So how did we end up with a single word, and why is that word "free"? Surprisingly, it comes from the same Old English root as "friend." According to etymologist Douglas Harper:

> [They both come] from the Old English *freon, freogan* "to free, love." The primary sense seems to have been "beloved, friend"; which in some languages (notably Gmc. and Celtic) developed a sense of "free," perhaps from the terms "beloved" or "friend" be-ing applied to the free members of one's clan (as opposed to slaves).
>
> The sense of "given without cost" is from 1585, from the notion of "free of cost."

So "free" comes from the social notion of freedom, both from slavery and from cost.

This book is about the "cost" meaning: free, as in beer. Or, for that matter, lunch.

A MILLION KINDS OF FREE

Even within the commercial use of "free" there is a wide range of meanings—and business models. Sometimes "free" isn't really free. "Buy one, get one free" is just another way of saying 50 percent off when you buy two. "Free gift inside" really means that the cost of the gift has been included in the overall product. "Free shipping" typically means the price of shipping has been built into the product's markup.

Of course, sometimes free really *is* free, but this is hardly a new economic model: A "free sample" is simple marketing, intended to both introduce a product and trigger a slight feeling of moral debt that may encourage you to buy the full-price item. A "free trial" may be free, but only for a limited time, and it may be difficult to opt out before it becomes paid. And "free air" at a gas station is what economists call a "complementary good"—a free product (DIY tire inflation) intended to reinforce consumer interest in a paid product (everything else at the gas station, from a pack of gum to the fuel).

Then there is the whole world of ad-supported media, from free-to-air radio and TV to most of the Web. Ad-supported free content is a business model that dates back more than a century: a third party (the

HOW CAN AIR TRAVEL BE FREE?

Every year, about 1.3 million passengers fly from London to Barcelona. A ticket on Dublin-based low-cost airline Ryanair is just $20 (£10). Other routes are similarly cheap, and Ryanair's CEO has said he hopes to one day offer all seats on his flights for free (perhaps offset by in-air gambling, turning his planes into flying casinos). How can a flight across the English Channel be cheaper than the cab ride to your hotel?

It costs Ryanair $70 to fly someone from London to Barcelona. Here is how it gets that money back:

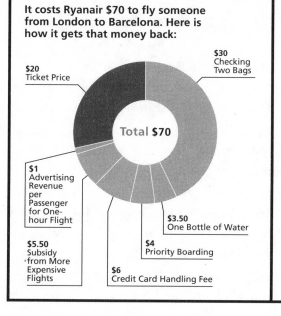

$20
Ticket Price

$30
Checking
Two Bags

Total $70

$1
Advertising
Revenue
per
Passenger
for One-
hour Flight

$5.50
Subsidy
from More
Expensive
Flights

$6
Credit Card Handling Fee

$4
Priority Boarding

$3.50
One Bottle of Water

▶ **Cut costs.** Ryanair boards and disembarks passengers from the tarmac to trim gate fees. The airline also negotiates lower access fees from less-popular airports eager for traffic.

▶ **Ramp up the ancillary fees.** Ryanair charges for in-flight food and beverages; assesses extra fees for preboarding, checked baggage, and flying with an infant; collects a share of car rentals and hotel reservations booked through the Web site; charges marketers for in-flight advertising; and levies a credit card handling fee for all ticket purchases.

▶ **Offset losses with higher fares.** On popular travel days, the same flight can cost more than $100.

advertisers) pays for a second party (the consumer) to get the content for free.

Finally, sometimes free really is free and *does* represent a new model. Most of this is online, where digital economics, with near-zero marginal costs, hold sway. Flickr, the photo-sharing service, is actually free for most of its users (it doesn't even use advertising). Likewise most of what Google offers either is free and without advertising or applies the media ad model in a new way to software and services (like Gmail), not content. Then there is the amazing "gift economy" of Wikipedia and the blogosphere, driven by the nonmonetary incentives of reputation, attention, expression, and the like.

All these can be sorted into four broad kinds of Free, two that are old but evolving and two that are emerging with the digital economy. Before we get to those, let's pull back and observe that all forms of Free boil down to variations of the same thing: shifting money around from product to product, person to person, between now and later, or into nonmonetary markets and back out again. Economists call these "cross-subsidies."

ALL THE WORLD'S A CROSS-SUBSIDY

Cross-subsidies are the essence of the phrase "there's no such thing as a free lunch." That means that one way or another the food must be paid for, if not by you directly then by someone else in whose interest it is to give you free food.

Sometimes people are paying indirectly for products. That free newspaper you're reading is supported by advertising, which is part of a retailer's marketing budget, which is built into its profit margin, which you (or someone around you) will ultimately pay for in the form of more expensive goods. You're also paying with a bit of your time and, by being seen reading that newspaper, your reputation. The free parking in the supermarket is paid for by the markup on the produce, and the free samples are subsidized by those who shell out for the paid versions.

HOW CAN A DVR BE FREE?

Phone companies sell calls; electronics companies sell gadgets. But cable giant Comcast is in both those businesses and a lot more besides. This gives it flexibility to cross-subsidize products, making one thing free in order to sell another. To that end, Comcast has given about 9 million subscribers free set-top digital video recorders. How can it make that money back?

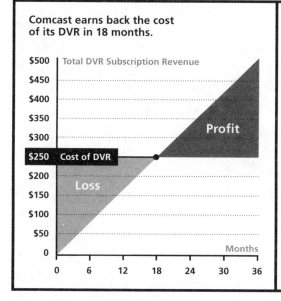

Comcast earns back the cost of its DVR in 18 months.

Total DVR Subscription Revenue

Profit

Cost of DVR

Loss

Months

▸ **Add hidden fees.** Comcast charges a $20 installation fee to every new DVR customer.

▸ **Charge a monthly subscription.** Comcast customers pay $14 a month to use the DVR box. Even if Comcast paid $250 for its DVRs — a very high estimate — the boxes would pay for themselves within 18 months.

▸ **Upsell other services.** Comcast hopes to win over customers with free DVRs, then interest them in services like high-speed Internet ($43 a month for 8 MBps) and digital telephony ($40 a month). That doesn't count pay-per-view movies, which can cost $5 each.

SOURCES: COMCAST, FORRESTER RESEARCH

In the gift economy (see page 27), the cross-subsidies are more subtle. Blogs are free and usually don't have ads, but that doesn't mean that value isn't being exchanged every time you visit. In return for the free content, the attention you give a blogger, whether in a visit or a link, enhances her reputation. She can use reputation to get a better job, enhance her network, or find more customers. Sometimes those reputation credits can turn into cash, but we can rarely predict the exact path—it's different each time.

Cross-subsidies can work in several different ways:

- **Paid products subsidizing free products.** Loss leaders are a staple of business, from the popcorn that subsidizes the loss-making movie to the expensive wine subsidizing the cheap meal in a restaurant. Free just takes that further, with one item being not just sold at a fraction of its cost but given away entirely. This can be as gimmicky as a "free gift inside" or as common as free samples. This form of Free is ancient, familiar, and relatively straightforward as an economic model, so we won't focus on it much here.

- **Paying later subsidizing free now.** The free cell phone with a two-year-subscription contract is a classic example of the subsidy over time. It's just shifting phone service from a point-of-sale revenue stream to an ongoing annuity. In this case, your future self is subsidizing your present self. The hope of the carrier is that you won't think about what you'll be paying each year for the phone service but instead will be dazzled by the free phone you get today.

- **Paying people subsidizing free people.** From the men who pay to get into nightclubs where the women get in free, to "kids get in free," to progressive taxation where the wealthy pay more so the less wealthy pay less (and sometimes nothing), the tactic of segmenting a market into groups based on their willingness or ability to pay is a conventional part of pricing theory. Free takes that to the extreme, extending the concept to a class of consumers who will get the product or service for nothing. The hope is that the free consumers will attract (in the case of the women) or bring with them (in the case of the kids) paying consumers or that some fraction of the free consumers will convert to paying consumers. When you walk through the striking interiors of Las Vegas attractions, you get the view for free; in exchange the owners are expecting some people to stop and gamble or shop (or, ideally, both).

Within the broad world of cross-subsidies, Free models tend to fall into four main categories:

FREE 1: DIRECT CROSS-SUBSIDIES

WHAT'S FREE: Any Product That Entices You
   to Pay for Something Else.
FREE TO WHOM: Everyone Willing to Pay Eventually,
   One Way or Another.

When Wal-Mart offers a buy-one-get-one-free deal on DVDs, it's a loss leader. The company is offering the DVD below cost to lure you into the store, where it hopes to sell you a washing machine or a shopping basket filled with other goods at a profit. In any package of products and services, from banking to mobile calling plans, the price of each individual component is often determined by psychology, not cost. Your cell phone company may not make money on your monthly minutes—it keeps that fee low because it knows that's the first thing you look at when picking a carrier—but your monthly voice mail fee is

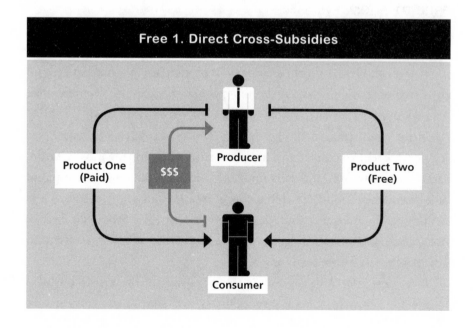

Free 1. Direct Cross-Subsidies

Product One (Paid) — $$$ — Producer — Product Two (Free) — Consumer pure profit. Companies look at a portfolio of products and price some at zero (or close to it) to make the other products, on which they make healthy profits, more attractive.

This is the extension, to more and more industries, of King Gillette's cross-subsidy. Technology is giving companies greater flexibility in how broadly they can define their markets, allowing them more freedom to give away some of their products or services to promote others. Ryanair, for instance, has disrupted its industry by defining itself more as a full-service travel agency than a seller of airline seats (see sidebar on page 19). Your credit card is free because the bank makes its money from the service charge it imposes on the retailers you buy from. They, in turn, pass that charge back to you. (Of course, if you don't pay your bill off in full at the end of the month, the bank makes even more money from your interest.)

FREE 2: THE THREE-PARTY MARKET

WHAT'S FREE: Content, Services, Software, and More.
FREE TO WHOM: Everyone.

The most common of the economies built around Free is the three-party system. Here a third party pays to participate in a market created by a free exchange between the first two parties. Sound complicated? You encounter it every day. It's the basis of virtually all media.

In the traditional media model, a publisher provides a product free (or nearly free) to consumers, and advertisers pay to ride along. Again, radio is "free to air," and so is much of television. Likewise, newspaper and magazine publishers don't charge readers anything close to the actual cost of creating, printing, and distributing their products. They're not selling papers and magazines to readers, they're selling readers to advertisers. It's a three-way market.

In a sense, the Web represents the extension of the media business model to industries of all sorts. This is not simply the notion that

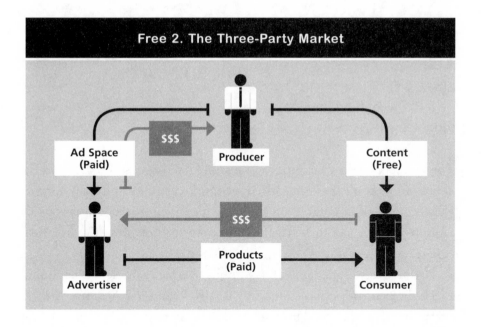

Free 2. The Three-Party Market

advertising will pay for everything. Media companies make money around free content in dozens of ways, from selling information about consumers to brand licensing, "value-added" subscriptions, and direct e-commerce (see Chapter 9 and the back of the book for a more complete list). Now an entire ecosystem of Web companies is growing up around the same set of models.

Economists call such models "two-sided markets," because there are two distinct user groups who synergistically support each other: Advertisers pay for media to reach consumers, who in turn support advertisers. Consumers ultimately pay, but only indirectly through the higher prices on products due to their marketing costs. This also applies to nonmedia markets, such as credit cards (free cards to consumers means more spending at merchants and more fees for issuing banks), operating system tools given free to application software developers to attract more consumers to the platform, and so on. In each case, the costs are distributed and/or hidden enough to make the primary goods feel free to consumers.

FREE 3: FREEMIUM

WHAT'S FREE: Anything That's Matched with
  a Premium Paid Version.
FREE TO WHOM: Basic Users.

This term, coined by venture capitalist Fred Wilson, is one of the most common Web business models. Freemium can take different forms: varying tiers of content from free to expensive, or a premium "pro" version of some site or software with more features than the free version (think Flickr and the $25-a-year Flickr Pro).

Again, this sounds familiar. Isn't it just the free sample model found everywhere from perfume counters to street corners? Yes, but with a pretty significant twist. The traditional free sample is the promotional candy bar handout or the diapers mailed to a new mother. Since these samples have real costs, the manufacturer gives away only a tiny

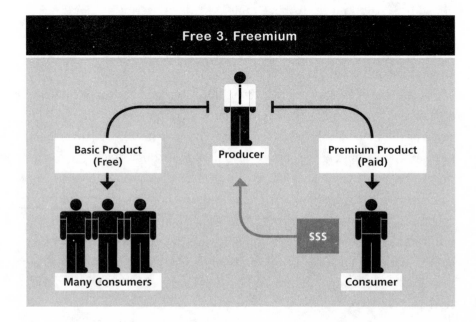

Free 3. Freemium

Basic Product (Free) · Producer · Premium Product (Paid)

Many Consumers · $$$ · Consumer

quantity—hoping to hook consumers and stimulate demand for many more.

But for digital products, this ratio of free to paid is reversed. A typical online site follows the 5 Percent Rule—5 percent of users support all the rest. In the freemium model, that means for every user who pays for the premium version of the site, nineteen others get the basic free version. The reason this works is that the cost of serving the nineteen is close enough to zero to call it nothing.

FREE 4: NONMONETARY MARKETS

WHAT'S FREE: Anything People Choose to Give Away
    with No Expectation of Payment.
FREE TO WHOM: Everyone.

This can take several forms:

Gift Economy

From the twelve million articles on Wikipedia to the millions of free secondhand goods offered on Freecycle (see sidebar on page 188), we are discovering that money isn't the only motivator. Altruism has always existed, but the Web gives it a platform where the actions of individuals can have global impact. In a sense, zero-cost distribution has turned sharing into an industry. From the point of view of the monetary economy it all looks free—indeed, it looks like unfair competition—but that says more about our shortsighted ways of measuring value than it does about the worth of what's created.

The incentives to share can range from reputation and attention to less measurable factors such as expression, fun, good karma, satisfaction, and simply self-interest (giving things away via Freecycle or Craigslist to save yourself the trouble of taking them to the dump). Sometimes the giving is unintentional, or passive. You give information to Google when you have a public Web site, whether you intend to or not, and you give aluminum cans to the homeless guy who collects them from the recycling bin, even if that's not what you meant to do.

Labor Exchange

You can get access to free porn if you solve a few Captchas, those scrambled text boxes used to block spam bots. Ironically, what you're actually doing is using your human pattern-matching skills to decipher text that originated on some other site, one of interest to spammers that uses such Captchas to keep them out. Once you solve it, the spammers can gain access to those sites, which are worth more to them than the bandwidth you'll consume viewing titillating images. As far as they're concerned, it's a black box—they put scrambled Captchas in and they get deciphered text out. But inside the box, it's the unwitting free labor of thousands of people. Likewise for rating stories on Digg, voting on Yahoo Answers, or using Google's 411 service (see sidebar on page 122). Every time you search on Google, you're helping the company improve its ad-targeting algorithms. In each case, the act of

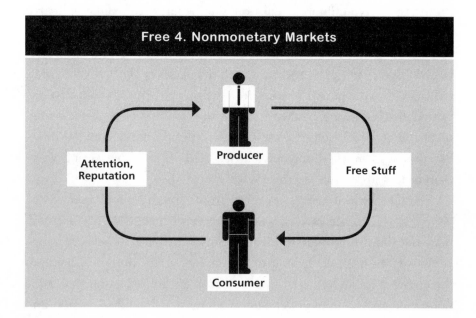

Free 4. Nonmonetary Markets

Attention, Reputation

Producer

Free Stuff

Consumer

using the service creates something of value, either improving the service itself or creating information that can be useful somewhere else. Whether you know it or not, you're paying with your labor for something free.

Piracy

This describes nothing so well as online music. Between digital reproduction and peer-to-peer distribution, the real cost of distributing music has truly hit bottom. This is a case where the product has become free because of sheer economic gravity, with or without a business model. That force is so powerful that laws, copy protection, guilt trips, and every other barrier to piracy the labels could think of failed (and continues to do so). Some artists give away their music online as a way of marketing concerts, merchandise, licensing, and other paid fare. But others have simply accepted that, for them, music is not a moneymaking business. It's something they do for other reasons, from fun to creative expression. Which, of course, has always been true for most musicians anyway.

A TEST OF FREE IN DAILY LIFE

Let's see how this taxonomy lines up with the sort of Free we encounter every day. Browsing a newsstand recently, I noticed a cover line on *Real Simple* magazine: "36 Surprising Things You Can Get for Free." It's the sort of thing you'll see on any newsstand in any month, so it seemed a fairly representative sample with which to test the framework. On page 30 is how the first half of *Real Simple*'s examples distributed.

You'll note that some of the examples have elements of several models, and others have competitors that use models that fit into different categories. (1-800-FREE411's competitor, Google 411, isn't ad-supported.) Also, government services are a special class of cross-subsidy, since the link between your taxes and the services you receive is indirect and diffuse.

A TEST OF FREE IN DAILY LIFE

Free Example	Free Model
	Free 1: Simple cross-subsidy
Apple Store classes	they're betting you'll buy something
Health club trials	ditto
Baby music classes	ditto
Ben and Jerry's Free Cone Day	ditto
Online photo printing (free samples)	ditto
Small business classes (government-funded)	you pay taxes
BBC language classes (podcasts)	cross-subsidy if you're British and pay taxes; gift economy if not
Popularity dialer (free excuse calls) 800-Free 411 Free reminder emails	**Free 2: Ad-supported**
	Free 3: Freemium (free & paid versions)
Skype (free phone calls)	(paid versions can connect to cell phones)
Kids night on Broadway	(parents support kids)
MIT OpenCourseWare (free classes online) Free pets on Craigslist Freecycle barter Museum (grants/donor funded) PaperbackSwap.com	**Free 4: Nonmonetary markets**

But the point holds: This sort of taxonomy works quite well. No category system is perfect, and it's not hard to find exceptions and hybrids, but this framework will serve us well in the chapters to come.

THE THREE PRICES

This book is mostly about two prices—*something* and *nothing*—but there is sometimes a third price that we can't ignore: *less than nothing.* That's right, a negative price: You get paid to use a product or service, rather than the other way around.

This is more common than you might think. Online, you can see this trend in things like Microsoft paying you to use their search, but it actually has a long tradition in conventional marketing. You find it in instant rebates and cash-back marketing, and in the cash rewards, frequent flyer miles, and other payments you get for using credit or loyalty cards.

Of course, few of these are really less than nothing; in most cases your wallet will open sooner or later. But what's interesting about these schemes is that although they're not really free money, consumers often treat them like they are.

For instance, a cash-back rebate invokes a very different psychology from simply saving the money in the first place. Studies of how people spend the $1,000 (or whatever amount) check they get when they buy a new truck (or, more to the point, finance it) show that they tend to spend it like a lottery winning—an unexpected windfall, even though it's really just a loan against future payments. Guys buy golf clubs their wives would never normally let them purchase, and their wives don't stand in their way, despite the fact that they *know* they'll be paying that money back over the years to come, just like a credit card debt.

In Dan Ariely's book *Predictably Irrational* there's a great example of negative pricing. In one instance, he told his class at MIT's Sloan School of Business that he would be doing a reading of poetry (Walt Whitman's *Leaves of Grass*) but didn't know what it should cost. He handed out a questionnaire to all the students, half of whom were asked if they'd be willing to pay $10 to hear him read, and the other half of whom were asked if they'd be willing to hear him read if he paid *them* $10. Then he gave them all the same question: What should the price be to hear him read short, medium, and long versions of the poem?

The initial question is what behavioral economists call an "anchor," which calibrates a consumer's sense of what a fair price is. It can have a dramatic effect on what they'll ultimately pay. In this case, the students who had been asked if they would pay $10 were willing to pay, on average, $1 for the short poem, $2 for the medium, and $3 for the long.

Meanwhile, the students who had been anchored to believe that Ariely should pay them did indeed demand that: They wanted $1.30 to listen to the short reading, $2.70 for the medium one, and $4.80 to endure the long reading.

Ariely notes that Mark Twain illustrated this with Tom Sawyer, who somehow got the other boys to be so envious of the fence-painting exercise that they not only took over his job but paid him for the privilege. However, there is a cautionary tale in this for those who would pay people for what they would otherwise expect to be paid themselves for. Twain observed: "There are wealthy gentlemen in England who drive four-horse passenger-coaches twenty or thirty miles on a daily line in the summer because the privilege costs them considerable money; but if they were offered a wage for the service, that would turn it into work and they would resign."

All these are examples of what Derek Sivers, the founder of CD Baby, calls "reversible business models." A real-world instance of this is the music clubs in Los Angeles that are charging bands to play in the club, rather than paying them as usual. The bands value the exposure more than the cash, and if they're good they can graduate to the usual sort of gigs.

In China, some doctors are paid monthly when their patients are healthy. If you are sick, it's their fault, so you don't have to pay that month. It's their goal to get you healthy and keep you healthy so they can get paid.

In Denmark, a gym offers a membership program where you pay nothing as long as you show up at least once a week. But miss a week and you have to pay full price for the month. The psychology is brilliant. When you go every week, you feel great about yourself and the gym. But eventually you'll get busy and miss a week. You'll pay, but you'll blame yourself alone. Unlike the usual situation where you pay for a gym you're not going to, your instinct is not to cancel your membership; instead it's to redouble your commitment.

FreeConferenceCall.com gets income from the phone companies instead of customers, because they know which phone company each person is using to call them. They negotiated an affiliate payment for generating more long-distance calls for each phone company. Rather than paying for the long-distance fees themselves, FreeConferenceCall charges the phone companies for encouraging users to make more long-distance calls.

Finally, Jicka.com, a company hoping to go up against Craigslist in free classifieds, decided to compete with Free by going one better: giving buyers trial warranties. If you place an ad to sell your house on Jicka, you'll get a six-month limited home warranty to give the buyer. Sell your car through a Jicka ad and the buyer will get a thirty-day limited warranty. Place any ad on Jicka and get one year of identity theft protection. This scheme doesn't cost Jicka anything. The companies that offer the warranties consider it a good marketing opportunity, and they'll make money when some of the Jicka free-trial customers decide to renew the warranties for a fee.

In each case, a clever company has reversed the normal flow of money, either making something free or paying for what other companies are charging for. There's nothing particularly high-tech about any of these ideas. They just took some entrepreneur thinking creatively about price.

3

THE HISTORY OF FREE

Zero, Lunch, and the Enemies of Capitalism

THE PROBLEM OF NOTHING

One of the reasons that Free is often so hard to grasp is that it is not a thing, but rather the absence of a thing. It is the hole where the price should be, the void at the till. We tend to think in terms of the concrete and tangible, yet Free is a concept, not something you can count on your fingers. It took thousands of years of civilization to even find a number to describe it.

The quantification of nothingness started, as so many things do, with the Babylonians. Around 3000 B.C., in the Fertile Crescent of present-day Iraq, a thriving agricultural society had a counting problem. It was not the obvious bug that you or I might have spotted, which is that their system was *sexagesimal,* or based on powers of sixties instead of tens. As awkward as that is, as long as you don't expect to count with your fingers and toes, it's easy enough to figure out (it is, after all, the root of our own time system).

No, the problem was something else: how to write down numbers.

Unlike most other cultures of that era, the Babylonians didn't have a different symbol for every number within their base set. Instead, they used just two marks: a wedge that represented 1 and a double wedge that represented 10. So, depending on where it was placed, a single wedge could represent 1; 60; 3,600; or an even greater multiple of sixty. It was, writes Charles Seife in *Zero: The Biography of a Dangerous Idea,* "the Bronze Age equivalent of computer code."

This made perfect sense in a culture that counted with an abacus. Adding numbers with that clever device is simply a matter of moving stones up and down, with stones in different columns representing different values. If you have abacuses with sixty stones in each column, a numbering system based on powers of sixty is no harder than one based on tens.

But when you want to mark a number on an abacus, what do you do if there are no stones in a column? The number 60 is one wedge in the sixties column and no wedges in the ones column. How do you write "no wedges"? The Babylonians needed a placeholder that represented nothing. They had to, in effect, invent zero. And so they created a new character, with no value, to signify an empty column. They denoted it with two slanted wedges.

Given the obvious need for such a placeholder when you're writing down numbers based on powers of any base, you might think that zero had been with us since the dawn of written history. But plenty of advanced civilizations managed to come and go with no need for it. The Romans had no use for it in Roman numerals. (There are no fixed columns in that notation. Instead, the value of any digit is determined by the other digits around it.)

The Greeks, meanwhile, explicitly rejected zero. Since their mathematical system was based on geometry, numbers had to represent space of one sort or another—length, angles, area, etc. Zero space didn't make sense. Greek math was epitomized by Pythagoras and his Pythagorean cult, which made such profound discoveries as the musical scale and the golden ratio (but not, ironically, the Pythagorean Theorem—the formula for calculating the hypotenuse of a right triangle had actually been known for many years before Pythagoras). Although they understood that arithmetic sometimes produces negative numbers, irrational numbers, and even zero, the Greeks rejected all of them because they could not be represented in physical shapes. (Awkwardly, the golden ratio is itself an irrational number, which was kept secret as long as possible.)

Such myopia is understandable. Where numbers only represent real things, you don't need a number to express the absence of something. It is an abstract concept and only shows up when the math gets equally abstract. "The point about zero is that we do not need to use it in the operations of daily life," wrote Alfred North Whitehead, the British mathematician, in 1911. "No one goes out to buy zero fish. It is in a way the most civilized of all the cardinal [numbers], and its use is only forced on us by the needs of cultivated modes of thought."

That fell to the mathematicians of India. Unlike the Greeks, the Indians did not see shapes in all numbers. Instead the Indians saw numbers as concepts. Eastern mysticism embraced both the tangible and the intangible, through the yin and yang of duality. The god Shiva was both the creator and the destroyer of worlds; indeed, one aspect of the deity Nishkala Shiva was the Shiva "without parts"—the void. Through their ability to divorce numerals from physical reality, the Indians invented algebra. That, in turn, allowed them to extend mathematics to its logical ends, including negative numbers and, by the ninth century, zero. Indeed, the very word "zero" has Indian origins: The Indian word for zero was *sunya*, meaning "empty," which the Arabs turned into *sifr*. Western scholars Latinized this into *zephirus*, the root of our zero.

THE PROBLEM OF FREE

By A.D. 900 there was both a symbol and an algebraic framework for nothing. But what about an economic system? Well, in a sense that had been there all along. The word "economics" comes from the Ancient Greek *oikos* ("house") and *nomos* ("custom" or "law"), therefore "rules of the house(hold)." And in the home, Free has always been the rule. Even after most cultures established monetary economies, day-to-day transactions within close-knit social groups, from families to tribes, was still mostly without price. The currencies of generosity, trust, goodwill, reputation, and equitable exchange still dominate the goods and services of the family, the neighborhood, and even within the workplace. In general, no cash is required among friends.

But for transactions between strangers, where social bonds are not the primary scoring system, money provided a common agreed-upon metric of value, and barter gave way to payment. But even then there was a place for Free, in everything from patronage to civil services.

As the nation-state emerged in the seventeenth century, so did the notion of progressive taxation, by which the rich gave more so the poor could pay less and receive services for free. This establishment of government institutions to serve the people created a special kind of Free: You may not pay for government services yourself, but society at large does, and you may never know exactly which of your own tax dollars come back to you directly.

Charity, of course, is also a form of Free, as is communal giving, such as barn raisings and potlatches, Native American gift festivals. The emergence of the five-day workweek, labor laws that established minimum and maximum work ages, and the shift from field labor to industrial and then white-collar work created free time. That, in turn, created a boom in volunteerism (free labor) that continues today.

Even as monetary economies became the norm, the importance of not charging for some things was still deeply held. Perhaps the best example is interest on a loan, which has historically been seen as a bit of an exploitation, especially when it comes to the poor. Today, "usury" means excessive interest, but it originally meant any interest whatsoever. (An interest-free loan is now seen as a form of gift.)

The early Catholic Church took a strong stand against charging for loans. In 1179, the Third Council of the Lateran decreed that persons who accepted interest on loans could receive neither the sacraments nor Christian burial. Pope Clement V made the belief in the right to usury heresy in 1311 and abolished all secular legislation that allowed it. Pope Sixtus V condemned the practice of charging interest as "detestable to God and man, damned by the sacred canons and contrary to Christian charity."

Not all societies saw interest as evil. The historian Paul Johnson notes:

> Most early religious systems in the ancient Near East, and the secular codes arising from them, did not forbid usury. These societies regarded inanimate matter as alive, like plants, animals and people, and capable of reproducing itself. Hence if you lent "food money," or monetary tokens of any kind, it was legitimate to charge interest. Food money in the shape of olives, dates, seeds or animals was lent out as early as c. 5000 B.C., if not earlier.

But when it comes to making a profit on hard cash, many societies have taken a hard stand. Some interpretations of Islamic law ban interest entirely, and the Koran minces no words on the subject:

> Those who charge usury are in the same position as those controlled by the devil's influence. This is because they claim that usury is the same as commerce. However, God permits commerce, and prohibits usury. Thus, whoever heeds this commandment from his Lord, and refrains from usury, he may keep his past earnings, and his judgment rests with God. As for those who persist in usury, they incur Hell, wherein they abide forever.

Eventually economic pragmatism made interest acceptable (and the Church came around, in part to appease the merchant classes to gain political support). In the sixteenth century, short-term interest rates dropped dramatically (from 20 to 30 percent annually to 9 to 10 percent), thanks to more efficient banking systems and commercial techniques, along with more money in circulation. The lower rates weakened most religious scruples about lending with interest, although Islamic law continued to frown on it.

CAPITALISM AND ITS ENEMIES

After the seventeenth century, the role of the market and the mercantile class became fully accepted pretty much everywhere. Money supplies were regulated, currencies were protected, and economies as we now know them flourished. More and more trade happened between strangers thanks to the principles of comparative advantage and specialization. (People made what they could make best and traded for other goods with people who could make them better.) Currencies became more important as the units of value because their worth came from trust in the overarching issuing authority (usually the state), rather than either of the parties in the transaction. The notion that "everything has its price" is just a few centuries old.

Thanks to Adam Smith, commerce became not just a place to shop but a way of thinking about all human activity. The social science of economics was born as a way to study why people make the choices they do. Just as in Darwin's description of nature, competition was at the heart of this emerging science of commerce. Money was how we kept score. Charging for things was simply the most efficient way to ensure that they would continue to be produced—the profit motive is as strong in economics as the "selfish gene" is in nature.

But amid the market triumphalism, there remained pockets of people who resisted money as the mediator of all exchange. Karl Marx advocated collective ownership and allocation according to need, not ability to pay. And the anarchist thinkers of the nineteenth century, such as the Russian prince-turned-radical Peter Kropotkin, imagined collectivist utopias where members would spontaneously perform all necessary labor because they would recognize the benefits of communal enterprise and mutual aid. Kropotkin believed that private property was one of the causes of oppression and exploitation and called for its abolition, advocating instead common ownership.

Spelling this out in his 1902 book, *Mutual Aid: A Factor of Evolution*, Kropotkin, in a way, anticipated some of the social forces that dominate the "link economy" of the Internet today (people linking to one another in their posts, bringing traffic and reputation to the recipient). In giving something away, he argued, the trade-off is not money, but satisfaction. This satisfaction was rooted in community, mutual aid, and support. The self-reinforcing qualities of that aid would, in turn, prompt others to give equally to you. "Primitive societies" worked that way, he argued, so such gift economies were closer to the natural state of human affairs than market capitalism.

But every effort to make this work in practice at any scale failed, largely because the social bonds that police such mutual aid tend to fray when the size of the group exceeds 150 (termed the "Dunbar number"—the empirically observed limit at which the members of a human community can maintain strong links with one another). Of course, this pretty much doomed collectivism for any group as large as a country. It would take the arrival of virtual worlds for us to finally see larger economies built on mutual benefit actually work. Online societies from the Web to online multiplayer games can allow us to maintain social networks that are much larger than those we maintain in the physical world. Software extends our reach and keeps score.

THE FIRST FREE LUNCH

By the end of the nineteenth century, it appeared that the ideological battles were largely over. Market economies were firmly established throughout the West. Far from the root of all evil, money was proving to be a catalyst of growth and the key to prosperity. The value of anything was best determined by the price people would pay for it—it was as simple as that. Utopian dreams of alternative systems based on gifts, barter, or social obligation were reserved for fringe experiments, from communes to Israel's kibbutzim. In the world of commerce, "free" took on its primary modern meaning: a marketing tool. And as such, it quickly became regarded with mistrust.

By the time King Gillette and Pearle Wait made their fortunes from Free, consumers were used to hearing "there's no such thing as a free lunch." The phrase refers to a tradition once common in U.S. saloons, which began offering "free" food to any customer who purchased at least one drink. Ranging from a sandwich to a multicourse meal, these free lunches were typically worth far more than the price of a single drink. However, the saloon-keepers were betting that most customers would buy more than one drink, and that the allure of free food would attract patrons during a less busy time of day.

In 1872, the *New York Times* reported that free lunches had emerged as a "peculiar" trend common in the Crescent City (New Orleans), where a free meal could be found in every saloon, every day. The spread included vast dishes of butter, large baskets of bread, huge vessels filled with potatoes, stewed mutton, stewed tomatoes, macaroni à la Français, and "a round of beef that must have weighed forty pounds."

According to the report, the free-lunch custom was feeding thousands of men who were subsisting "entirely on meals this way." The *Times* article continued:

A free-lunch counter is a great leveler of classes, and when a man takes a position before one of them he must give up all hope of appearing dignified. . . . All classes of the people can be seen partaking of these free meals and pushing and scrambling to be helped a second time.

In fact, the nearly indigent "free-lunch fiend" actually became a recognized social type. Another *New York Times* story from 1872 cites free-lunch loafers who "toil not, yet they get along," visiting saloons to try to bum drinks from strangers. "Should this inexplicable lunch-fiend not be called to drink, he devours whatever he can, and while the bartender is occupied, attempts to escape unnoticed."

In San Francisco the custom arrived with the Gold Rush and stayed for years. An 1886 *Times* article on the fading of the days of the forty-niners in San Francisco calls the free-lunch fiend the "only landmark of the past." The story begins "How do all these idle people live?" and then assigns responsibility. "Take away that peculiarly California institution," the piece concludes, "and they would starve."

Elsewhere, the free lunch ran afoul of the temperance movement. An 1874 history of the battle to ban alcohol suggests that the free lunch—along with women and song—was nothing but a way to disguise a well-filled bar. The alcohol was the "centre about which all these other things are made to revolve."

Others argued that the free lunch actually performed a social-relief function. Reformer William T. Stead commented that in the winter of 1894 the free-lunch saloons "fed more hungry people in Chicago than all the other agencies, religious, charitable, and municipal, put together." Stead cites a newspaper's estimate that the saloon-keepers fed sixty thousand people a day. And, he noted, those three thousand saloons did so without pretending to be charitable.

Two years later, in 1896, the New York State Legislature passed the Raines Law, which was intended to regulate liquor traffic. Among its many provisions, one forbade the sale of liquor unless the purchase was accompanied by food. Another outlawed the free lunch altogether. That provision was short-lived, though. The following year, the Raines Law was amended to permit free lunches. The custom continued for decades more.

SAMPLES, GIFTS, AND TASTERS

At the beginning of the twentieth century, Free re-emerged along with the new packaged goods industry. With the rise of brands, advertising, and national distribution, Free became a sales gimmick. There's nothing new about free samples, but the mass marketing of them is credited to a nineteenth-century marketing genius named Benjamin Babbitt.

Among Babbitt's many inventions were several methods for making soap. But where he really shined was in his innovative selling, which rivaled even that of his friend P. T. Barnum. Babbitt's Soap became nationally famous due to his advertising and promotional campaigns, which included the first widespread distribution of free samples. "A fair trial is all I ask for," his advertisements proclaimed, showing gentlemen salesmen giving away samplers.

The soap was sold from brightly painted streetcars with musicians, which led to the phrase "get on the bandwagon." In 1922, Sinclair Lewis used the Babbitt name as the title character of his best seller, *Babbitt,* about a vulgar and ignorant businessman.

Another pioneering example is Wall Drug in South Dakota. In 1931, Ted Hustead, a Nebraska native and pharmacist, was looking to establish his business in a small town with a Catholic church. He found exactly what he wanted with Wall Drug. It was located in a 231-person town in what he referred to as "the middle of nowhere." Understandably, the store struggled. But in 1933, Mount Rushmore opened sixty miles to the west, and Hustead's wife, Dorothy, got the idea to advertise free ice water to parched travelers heading to see the monument. The tactic put Wall Drug on the map, and business boomed.

Today Wall Drug is an enormous cowboy-themed shopping mall/department store. It now offers free bumper stickers and free promotional signs, along with 5-cent coffee. Ice water, of course, is still free.

FREE AS A WEAPON

One of the first hints of the twenty-first-century power of Free came at the dawn of the transformative medium of the twentieth century—radio. Today, we know that the most disruptive way to enter a market is to vaporize the economics of existing business models. Charge nothing for a product that the incumbents depend on for their profits. The world will beat a path to your door and you can then sell them something else. Just look at free long-distance calling with mobile phones, which decimated the fixed line long-distance business, or think of what free classifieds do to newspapers.

Seventy years ago, a similar battle played out over recorded music. In the late 1930s, radio was emerging as a popular entertainment format, but also one that made a mess of the old ways of paying musicians. Most radio music broadcasts of the time were live, and the musicians and composers were paid for a single performance. But to the artists, payment for a single performance alone didn't seem fair when that one performance was being received by millions of listeners. Had those millions been packed into one concert hall, the musicians' share of the receipts would have been much larger.

Broadcasters argued that it was impossible to pay licensing fees based on how many listeners tuned in, because no one knew what that number was. But ASCAP, with its near-monopoly on the most popular artists, made the rules: It insisted on royalties of 3 to 5 percent of a station's gross advertising revenues in exchange for the right to play music. Worse, it threatened to raise that rate when the contract expired in 1940.

As the broadcasters and ASCAP were negotiating, radio stations started taking matters into their own hands, and cut the live performances out entirely. Recording technology was improving, and more and more stations began playing records, which were introduced by a studio announcer known as a disk jockey. The music labels responded by selling records stamped with "NOT LICENSED FOR RADIO BROADCAST," but in 1940 the Supreme Court ruled that radio stations, having purchased a record, could play it. So ASCAP convinced its most prominent members, such as Bing Crosby, to simply halt making new recordings.

Faced with a shrinking pool of music to play and a potentially ruinous royalty requirement, the broadcasters struck back by organizing their own royalty agency, Broadcast Music Incorporated (BMI). The upstart BMI quickly became a magnet for regional musicians, such as rhythm-and-blues and country-and-western artists, who were traditionally neglected by the New York–based ASCAP. Because these less popular musicians wanted exposure more than money, they agreed to let the radio stations broadcast music for free. The business model of charging radio stations a fortune for the right to play music collapsed. Instead, radio was recognized as a prime marketing channel for artists, who would make their money from selling records and concerts.

Although ASCAP challenged this in several lawsuits in the 1950s and 1960s, it never regained the power to charge high royalties to radio stations. Free-to-air radio plus nominal royalties for artists created the disk jockey era and, in turn, the Top 40 phenomenon. Today these royalties are calculated based on a formula involving time, reach, and type of station, but are low enough for radio stations to prosper.

The irony was complete. Rather than undermining the music business, as ASCAP had feared, Free helped the music industry grow huge and profitable. A free inferior version of the music (lower quality, unpredictable availability) turned out to be great marketing for a paid superior version, and the artists' revenues shifted from performance to record royalties. Now Free offers the opportunity to switch back again, as free music serves as marketing for the growing concert business. The one constant, predictably, is that the labels are still against it.

THE AGE OF ABUNDANCE

If the twentieth century saw people starting to embrace Free again as a concept, it also witnessed a crucial phenomenon that helped to make Free a reality—the arrival of abundance. For most previous generations, scarcity—of food, of clothing, or of shelter—was a constant concern. For those born in the developed world in the past half century or so, however, abundance has been the keynote. And nowhere has that abundance been more apparent than in that fundamental prerequisite for life: food.

When I was a kid, hunger was one of the main problems of poverty in America. Today, it's obesity. Something dramatic has changed in the world of agriculture in the past four decades—we got much better at growing food. A technology-driven revolution turned a scarce commodity into an abundant one. And in that story lie clues to what can happen when any major resource shifts from scarcity to abundance.

There are only five major inputs to a crop: sun, air, water, land (nutrients), and labor. Sun and air are free, and if the crop is grown in an area with plenty of rainfall, water can be free, too. The remaining inputs—primarily labor, land, and fertilizer—are very much not free, and they account for most of the price of crops.

In the nineteenth century, the Industrial Revolution mechanized agriculture, radically lowering the cost of labor and increasing crop yield. But it was the "Green Revolution" of the 1960s that really transformed the economics of food by making farming so efficient that fewer people had to do it anymore. The secret of this second revolution was chemistry.

For most of human history manure has determined how much food we had. Agricultural yield was limited by the availability of fertilizer, and that largely came from animal (and sometimes human) waste. If a farm wanted to support both animals and crops in a synergistic nutrient cycle, it had to split its land between them. But at the end of the nineteenth century, naturalists began to understand what it was in manure that plants need: nitrogen, phosphorous, and potassium.

At the beginning of the twentieth century, a few chemists started work on making those elements synthetically. The breakthrough came when Fritz Haber, working for BASF, figured out how to extract nitrogen from the air in the form of ammonia by combining air with natural gas under high pressure and heat. Commercialized by Carl Bosch in 1910, cheap nitrogenous fertilizer—along with the sudden availability of it—is credited with averting the long-predicted "Malthusian catastrophe," or population crisis. Today, production of ammonia currently constitutes about 5 percent of global natural gas consumption, accounting for around 2 percent of world energy production.

The Haber-Bosch Process eliminated farmers' dependency on manure. Along with chemical pesticides and herbicides, this created the Green Revolution, which increased agricultural capacity worldwide nearly a hundredfold, allowing the planet to feed a growing population, especially a new middle class that, desiring to eat higher on the food chain, increasingly chose resource-intensive meat rather than just grains.

The effects of this have been dramatic. The cost of feeding ourselves has dropped from one-third of the average U.S. household income in 1955 to less than 15 percent today.

PILING CORN UPON CORN

One aspect of agricultural abundance that touches every one of us every day is the Corn Economy. This extraordinary grass, bred by man over millennia to have larger and larger starch-filled kernels, produces more food per acre than any other plant on the Earth.

Corn economies are naturally abundant economies, at least as far as food goes. Historians often look at the great civilizations of the ancient world through the lens of three grains: rice, wheat, and corn. Rice is protein-rich but extremely hard to grow. Wheat is easy to grow but protein-poor. Only corn is both easy to grow and plump with protein.

What historians have observed is that the protein/labor ratio of these grains influenced the course of the civilizations based on them. The higher that ratio, the more "social surplus" the people eating that grain had, since they could feed themselves with less work. The effect of this was not always positive. Rice and wheat societies tended to be agrarian, inwardly focused cultures, presumably because the process of raising the crops took so much of their energy. But corn cultures—the Mayans, the Aztecs—had spare time and energy, which they often used to attack their neighbors. By this analysis, corn's abundance made the Aztecs warlike.

Today, we use corn for more than just food. Between synthetic fertilizer and breeding techniques that make corn the most efficient converter of sunlight and water to starch the world has ever seen, we are now swimming in a golden harvest of plenty—far more than we can eat. So corn has become an industrial feedstock for products of all sorts, from paint to packaging. Cheap corn has driven out many other foods from our diet and converted natural grass-eating animals, such as cows, into corn-processing machines.

As Michael Pollan points out in *The Ominivore's Dilemma,* a chicken nugget "piles corn upon corn: what chicken it contains consists of corn [its feed], but so do the nugget's other constituents, including the modified corn starch that glues the thing together, the corn flour in the batter and the corn oil in which it is fried. Much less obviously, the leavenings and the lecithin, the mono-, di- and triglycerides, the attractive golden color and even the citric acid that keeps the nugget fresh can all be derived from corn."

A quarter of all the products found in an average supermarket today contain corn, Pollan writes. And that goes for the nonfood items, too! From toothpaste and cosmetics to disposable diapers and cleansers, everything contains corn, even the cardboard they're boxed in. Even the supermarket itself, with its wallboard and joint compound, linoleum and adhesives, is built on corn.

Corn is so plentiful that we now use it to make fuel for our cars, in the form of ethanol, which has finally tested its abundance limits. After decades of price declines, corn has in recent years started getting more expensive along with oil prices. But innovation abhors a rising commodity, so that rising price has simply accelerated the search for a way to make ethanol out of switchgrass or other forms of cellulose, which can be grown where corn cannot. Once that magic cellulose-eating enzyme is found, corn will get cheap again, and with it, food of all sorts.

EHRLICH'S BAD BET

The idea that commodities might get cheaper, not more expensive, over time is counterintuitive. Food is at least replenishable, but minerals are not. After all, the Earth is a limited resource, and the more ore we take out of it the less there remains, which is a classic case of scarcity. In 1972, a think tank called the Club of Rome published a book called *Limits to Growth* that modeled the consequences of a rapidly growing world population and finite resource supplies. It went on to sell 30 million copies and defined the environmental movement, including the dangers of the "population bomb" that was putting a higher burden on our planet than it could conceivably take.

But not everyone agreed with this Malthusian despair. A look at the history of the nineteenth and twentieth centuries suggested that we get smarter faster than we reproduce—human ingenuity tends to find ways to extract resources from the earth faster than we can use them. This has the effect of increasing supply faster than demand, which in turn depresses prices. (Obviously this can't go on forever, since those resources are ultimately limited, but the point was that they are a lot less limited than the Club of Rome thought.) The debate surrounding the veracity of this statement turned into one of the most famous bets in history, one that would essentially define the opposing views of scarcity versus abundance thinking.

In September 1980, Paul Ehrlich, a population biologist, and Julian Simon, an economist, made a wager, publicly recorded in the pages of *Social Science Quarterly,* over the future price of some core commodities.

Simon made a public offer to stake $10,000 on his belief that "the cost of non-government-controlled raw materials (including grain and oil) will not rise in the long run." Ehrlich took the bet, and they designated September 29, 1990, ten years hence, as the payoff date. If the inflation-adjusted prices of various metals rose over that period, Simon would pay Ehrlich the combined difference; if the prices fell, Ehrlich would pay Simon. Ehrlich chose five metals: copper, chrome, nickel, tin, and tungsten.

Between 1980 and 1990, the world's population grew by more than 800 million, the largest increase in one decade in all of history. But by September 1990, without a single exception, the price of each of Ehrlich's selected metals had fallen, and in some cases by more than half. Chrome, which had sold for $3.90 a pound in 1980, was down to $3.70 in 1990. Tin, which was $8.72 a pound in 1980, was down to $3.88 a decade later.

Why did Simon win the bet? Partly because he was a good economist and understood the substitution effect: If a resource becomes too scarce and expensive, it provides an incentive to look for an abundant replacement, which shifts demand away from the scarce resource (witness the current race to find replacements for oil). Simon believed—rightly so— that human ingenuity and the learning curve of science and technology would tend to create new resources faster than we used them.

He also won because Ehrlich was simply too pessimistic. Ehrlich had predicted famines of "unbelievable proportions" occurring by 1975, with hundreds of millions of people starving to death in the 1970s and 1980s, which would signify that the world was "entering a genuine age of scarcity." (Despite his miscalculations, Ehrlich received a MacArthur Foundation Genius Award in 1990 for having promoted "greater public understanding of environmental problems.")

Humans are wired to understand scarcity better than abundance. Just as we've evolved to overreact to threats and danger, one of our survival tactics is to focus on the risk that supplies are going to run out. Abundance, from an evolutionary perspective, resolves itself, while scarcity needs to be fought over. The result is that despite Simon's victory, the world seemed to assume that Ehrlich, on some level, was still right.

Simon complained that, for some reason he could never comprehend, people were inclined to believe the very worst about anything and everything; they were immune to contrary evidence just as if they'd been medically vaccinated against the force of fact. Ehrlich's gloomy predictions continued (and continue) to have influence. Meanwhile Simon's own observations seem to be of interest only to commodities traders.

CORNUCOPIA BLINDNESS

It should have been obvious that Simon stood a better chance of winning the bet. But our tendency to give scarcity more attention than abundance has caused us to ignore the many examples of abundance that have arisen in our own lifetime, like corn, for starters. The problem is that once something becomes abundant, we tend to ignore it, just like we ignore the air that we breathe. There is a reason why economics is defined as the science of "choice under scarcity": In abundance you don't have to make choices, which means that you don't have to think about it at all.

You can see this in examples small and large. In some landlocked parts of Europe during the Middle Ages salt was at times so scarce that it was used as a "currency" like gold. Just look at it now: It's a condiment included free with any meal—too cheap to meter.

In the broader category there are sweeping effects such as globalization, which made abundant labor available to any country. Today basic necessities such as clothing can be made so cheaply as to be essentially disposable. In 1900, the most basic man's shirt (essentially the fabric and sewing equivalent of a T-shirt) in the United States cost about $1 wholesale, which was a lot, especially after it was marked up for retail. As a result, the average American consumer had just eight outfits.

Today, that T-shirt still costs $1 wholesale. But $1 today is worth one-twenty-fifth what $1 was worth a century ago, which means that in practice we can buy twenty-five shirts for the price of one from back then. There is no need for anyone to dress in rags today; indeed some homeless have easier access to free clothing than they do to showers and washing machines, so they simply treat clothing as a disposable item, to be worn for a short while and then discarded.

But perhaps the most familiar example of abundance in the twentieth century was plastic, which made atoms almost as costless and malleable as bits. What plastic, the ultimately fungible commodity, could do was to reduce manufacturing and material costs to practically nothing. It didn't need to be carved, machined, painted, cast, or stamped. It was simply molded in any shape, texture, or color desired. The result was the birth of disposable culture. The concept King Gillette introduced with the razor blade was extended to nearly everything else by Leo Baekeland, who created the first all-synthetic polymer in 1907. His name gave us Bakelite. As embodied by the company's logo, the letter B hovering over the mathematical symbol for infinity, the polymer's applications seemed endless.

In World War II plastic became a key strategic material and the U.S. government spent a billion dollars on synthetic polymer production plants. After the war, all this capacity, redirected to the consumer market, turned a remarkably malleable material into an exceedingly cheap one. And thus were born Tupperware, Formica tables, fiberglass chairs, Naugahyde love seats, hula hoops, disposable pens, Silly Putty, and nylon panty hose.

The first generation of plastic was sold not as a disposable substance but rather as a superior one. It could be formed into more perfect shapes than metal and was more lasting than wood. But the second generation of plastics, the vinyls and polystyrenes, were so cheap that they could be tossed out without a thought. In the 1960s, brightly colored disposable goods represented modernity, the triumph of industrial technology over material scarcity. Throwing away manufactured goods was not wasteful; it was the privilege of an advanced civilization.

After the 1970s, attitudes toward this superabundance began to change. The environmental cost of a disposable consumer culture became more obvious. Plastic may have seemed close to free, but that's only because we weren't pricing it properly. Include the environmental costs—the "negative externalities"—and maybe it doesn't feel as right to toss out that McDonald's Happy Meal toy after one play. A generation started recycling. Our attitudes toward abundant resources moved from personal psychology ("it's free to me") to collective psychology ("it's not free to us").

ABUNDANCE WINS

The story of the twentieth century is extraordinary social and economic change driven by abundance. The automobile was enabled by the ability to tap vast stores of petroleum, which replaced scarce whale oil and made liquid fuels ubiquitous. The eighty-foot container, which didn't need a dock full of longshoremen to load and unload, made shipping cheap enough to tap abundant labor far away. Computers made information abundant.

Just as water will always flow downhill, economies flow toward abundance. Products that can become commoditized and cheap tend to do so, and companies seeking profits move upstream in search of new scarcities. Where abundance drives the costs of something to the floor, value shifts to adjacent levels, something technology publisher Tim O'Reilly calls the "Law of Conservation of Attractive Profits."

In 2001, management guru Seth Godin wrote in *Unleashing the Ideavirus,* "Twenty years ago, the top 100 companies in the Fortune 500 either dug something out of the ground or turned a natural resource (iron ore or oil) into something you could hold." Today, as Godin observed, it's very different.

Only thirty-two of the Top 100 companies today make things you can hold, from aerospace and motor vehicles to chemicals and food, metal bending and heavy industry. The other sixty-eight traffic mostly in *ideas*, not resource processing. Some offer services rather than goods, such as health care and telecommunications. Others create goods that are mostly intellectual property, such as drugs and semiconductors, where the cost to produce the physical product is tiny compared to the cost of inventing it. Yet others create markets for other people's goods, such as mass retailers and wholesalers. Here's a breakdown of the list:

- Insurance: Life, Health (12)
- Health Care (6)
- Commercial Banks (5)
- Wholesalers (5)
- Food and Drug Stores (5)
- General Merchandisers (4)
- Pharmaceuticals (4)
- Securities (4)
- Specialty Retailers (4)
- Telecommunications (4)
- Computers, Office Equipment (3)
- Entertainment (3)
- Diversified Financials (2)
- Mail, Package, Freight Delivery (2)
- Network and Other Communications Equipment (2)
- Computer Software (1)
- Savings Institutions (1)
- Semiconductors and Other Electronic Components (1)

The point, which we learned from the Ehrlich/Simon bet, was that as commodities become cheaper, value moves elsewhere. There's still a lot of money in commodities (witness the oil-producing states), but the highest profit margins are usually found where gray matter has been added to things. That's what happened to the above list. A few decades ago, the most value was in manufacturing. Then globalization rendered manufacturing a commodity, and the price fell. So the value moved to things that were not (yet) commodities, further away from hand-eye co-ordination and closer to brain-mouth coordination. Today's knowledge workers are yesterday's factory workers (and the day before's farmers) moving upstream in search of scarcity.

These days that scarcity is what former U.S. labor secretary Robert Reich called "symbolic analysis," the combination of knowledge, skills, and abstract thinking that defines an effective knowledge worker. The constant challenge is to figure out how best to divide labor between people and computers, and that line is always moving.

As computers are taught to do a human job (like stock trading), the price of that job drops closer to zero, and the displaced humans either learn to do something more challenging or they don't. The first group typically gets paid more than they used to and the second group gets paid less. The first is the opportunity that comes with industries moving toward abundance; the second is the cost. As a society, our job is to try to make the first group bigger than the second.

Abundance thinking is not only discovering what will become cheaper, but also looking for what will become more valuable as a result of that shift, and moving to that. It's the engine of growth, something we've been riding since even before David Ricardo defined the "comparative advantage" of one country over another in the eighteenth century. Yesterday's abundance consisted of products from another country with more plentiful resources or cheaper labor. Today's also consists of products from the land of silicon and glass threads.

4

THE PSYCHOLOGY
OF FREE

It Feels Good. Too Good?

IN 1996, the *Village Voice* finally gave in. Forty years after its founding, the legendary publication stopped charging a cover price. Like almost all other weekly city newspapers, it would become free, distributed in boxes on the street and in stacks at friendly retailers. This was pretty much universally marked as the day the *Village Voice* stopped mattering. A 2005 profile of the paper in *New York* magazine was headlined "The Voice from Beyond the Grave: The legendary downtown paper has been a shell of its former self since it went free nearly a decade ago."

Now contrast that with *The Onion,* another weekly newspaper. Started in 1988 as a free satirical broadsheet in the college town of Madison, Wisconsin, *The Onion* has grown into an empire. Over the past two decades, it has expanded its regional print editions in ten other cities and launched a Web site that now gets millions of visitors each month. It publishes books, produces a TV show, and dabbles in feature-length movies. *The Onion* was born free, stayed free, and continues to thrive.

On the face of it, the story of these two publications is puzzling.

Free seemingly killed one weekly newspaper but animated another. In one case, Free devalued the product, while in another it drove an impressive expansion.

But it's not as simple as that. For starters, Free didn't cause the demise of the *Village Voice*. As the *New York* article explained:

> Told that many writers felt that the impact of their work had been diminished when the paper went free, [publisher] David Schneiderman scoffed, adding that there was no choice. "We were below 130,000 circulation, down from a top of 160,000. Now the circulation is 250,000. . . . Wouldn't you rather be read by twice as many people?" . . . It wasn't going free that hurt the paper. It *saved* the paper. Kept it going, making money.

In other words, the *Voice* had been in decline, at least in terms of its business fundamentals, for many years before it went free; people confused cause and effect.

Why do people think "free" means diminished quality in one instance, and not in another? It turns out that our feelings about "free" are relative, not absolute. If something used to cost money and now doesn't, we tend to correlate that with a decline in quality. But if something *never* cost money, we don't feel the same way. A free bagel is probably stale, but free ketchup in a restaurant is fine. Nobody thinks that Google is an inferior search engine because it doesn't charge.

With *The Onion* and the *Village Voice,* we get at one crucial misconception about Free, but only in the context of two prices—zero and non-zero. In today's media marketplace, the psychology of Free (and therefore pricing) is actually a bit more nuanced. Let me give you an example that is closer to home: a glossy monthly magazine. It can typically be obtained in several different ways. You can read it online for free, in a somewhat stripped-down form that trades the design and photography packaging of the print edition (which is hard to re-create on the Web) for instant accessibility. Or you can buy one issue of the magazine on the newsstand for, say, $4.95. Or you can subscribe and get a year (twelve issues) for as little as $10, which is 83 cents per issue, delivered right to your door. Where do those three prices—$0, $4.95, and $0.83—come from?

The Web price (free) is the easiest. The cost of delivering the content is so low that publishers round down to zero and use Free to reach the largest possible audience. They may put an average of two ads on every page, each of which is sold for between $5 and $20 per thousand views. That means they get between 1 and 4 cents of revenue for each page someone looks at. The cost of serving that page is only a fraction of a cent. (The rest of the costs are in creating the content in the first place, but publishers amortize that over the entire audience: The bigger the readership, the lower the cost per page.)

The next simplest price is the newsstand price of $4.95. The newsstand keeps less than half, to pay their costs and make a profit. The rest goes to the publisher and provides a dollar or two of profit after the costs of printing and distribution. But for most magazines, more than half the copies they print don't actually sell, which means they are returned and pulped. That can cut the profit considerably. So why bother with the newsstand? Because it's a good way to acquire new subscribers, since they can sample the real thing rather than just read a letter describing it. Plus, publishers can make a decent profit from the advertising in the copies that do sell.

So far those prices are set by economics, not psychology. But what about the $10 yearly subscription? Well, here's where it gets interesting. The actual cost of printing and mailing twelve issues to your home is $15, and when you add the cost of acquiring you as a subscriber in the first place, that can add up to more than $30 per year per subscriber. And yet they charge just $10. There's no magic at work here. The advertising makes up the difference, so that $10 of direct revenues from the subscriber is topped up by the advertiser. Advertising makes a loss-leading subscription model profitable. And if the subscriber stays for three years or more, even the acquisition costs are repaid, making them more profitable yet.

But why $10? If the publisher is able to subsidize its subscribers by more than 60 percent, surely it could go all the way to 100 percent and make the subscription free? Ah, now we're getting into psychology.

The simple answer is that the act of writing a check or entering a credit card number, *regardless of the amount,* is an act of consumer volition that completely changes how an advertiser sees a reader. Writing a check for any amount (even 1 cent) means that you actually want the magazine, and will presumably read it and treasure it when it arrives. In fact, advertisers will pay as much as *five times more* to be part of that relationship than they'll pay for a free magazine that may be treated as junk mail.

However, there are plenty of magazines that do give away free subscriptions. That's called "controlled circulation" and it's based on another currency: information. These magazines tend to be very focused business periodicals, such as those aimed at chief financial officers or others with corporate purchasing power, or targeted "tastemaker" lifestyle magazines.

These business magazines' readers certify—well, claim—that they are important people with huge wads of cash to spend, and the magazine can use this information to charge the advertisers higher rates to reach them. In this case, having a lot of desirable executives on their subscription rolls, each of whom has nominally filled out a form claiming to want the magazine, compensates in the eyes of the advertisers for the fact that these readers have not put any actual money where their mouth is. A similar type of focused circulation has also been successful for *Vice,* an irreverent lifestyle magazine aimed at twentysomethings. Freely distributed at hip coffee shops, record stores, and clothing boutiques—initially in Canada in the 1990s, then the United States, then worldwide—*Vice* gave advertisers access to an influential audience they might not otherwise reach. The small print publication eventually grew into a record label, a retail clothing chain, Vice Film, and VBS.tv, a Web television venture.

Okay, so that explains why most publishers don't give away subscriptions for free. But how did they arrive at $10? That price is all about perception. It is the lowest sum that is not too low to devalue the product. Lower is better for subscribers, since the less they have to pay, the more likely they are to sign up. But higher is better for advertisers, because the more a consumer pays for a product, the more they value it. So $10 is low enough to get a lot of people to subscribe, while not being so low that it discredits the product in the eyes of the advertisers. (That same devaluation of something very cheap can also affect how subscribers feel, but we can't measure it as well as we can the advertiser reaction.)

THE PENNY GAP

With magazines it can clearly be effective to charge a minimal price, instead of nothing. But in most cases, just a penny—a seemingly inconsequential price—can stop the vast majority of consumers in their tracks. A single penny doesn't really mean anything to us economically. So why does it have so much impact?

The answer is that it makes us think about the choice. That alone is a disincentive to continue. It's as if our brains were wired to raise a flag every time we're confronted with a price. This is the "is it worth it?" flag. If you charge a price, any price, we are forced to ask ourselves if we really want to open our wallets. But if the price is zero, that flag never goes up and the decision just got easier.

The proper name for that flag is what George Washington University economist Nick Szabo has dubbed "mental transaction costs." These are, simply, the toll of thinking. We're all a bit lazy and we'd rather not think about things if we don't have to. So we tend to choose things that require the least thinking.

The phrase "transaction costs" has its roots in the theory of the firm, Nobel Prize–winning economist Ronald Coase's explanation that companies exist to minimize the communications overhead within and between teams. This refers mostly to the cognitive load of having to process information—figuring out who should do what, whom to trust, and the like.

Szabo extended this to purchasing decisions. He looked at the idea of "micropayments," financial systems that would allow you to pay fractions of a cent per Web page you read, or millieuros for each comic strip you download. All these schemes are destined to fail, Szabo concluded, because although they minimize the economic costs of choices, they still have all the cognitive costs.

For example, consider a PowerPoint presentation on "ten time-saving ideas for a penny each." The mental energy of deciding if the whole thing is worth 10 cents, or if each individual idea is worth a penny, just isn't worth it. Many potential customers would be put off by the payment and decision process. Meanwhile the revenues generated by such micropayments are, by definition, tiny. It's the worst of both worlds—the mental tax of a larger price without the commensurate cash. (Szabo was right: Micropayments have largely failed to take off.)

HOW CAN EVERYTHING IN A STORE BE FREE?

At SampleLab, a boutique in Tokyo's teen-laden Harajuku district, customers get up to five free items each time they visit — everything from candles, noodles, and face cream to the occasional $50 videogame cartridge. The gratis-only "sample salon" attracts 700 visitors a day. How can SampleLab not charge for every item it stocks?

Most monthly revenue comes from selling shelf space and customer feedback.

■ SHELF RENTAL
■ SURVEY DATA
■ MEMBERSHIP FEES

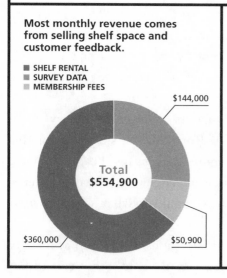

$144,000

Total
$554,900

$360,000

$50,900

▶ **Charge for entry.** Only "members," who pay $13 in registration and annual fees, are admitted. With 47,000 members, SampleLab is so hip, teens now have to make reservations one week in advance.

▶ **Charge a "rental" fee for shelf space.** Due to the store's popularity, companies give SampleLab products for free and even pay $2,000 to stock one item for two weeks. SampleLab can carry 90 products at once.

▶ **Charge for feedback.** By offering extra free goods, SampleLab turns most of its members into a focus group. Teens fill out product-specific surveys online, on paper, or via *keitai* (cell phone). Companies pay $4,000 for the data. If 20 percent of its clients pay for the feedback, SampleLab earns a little less than half the monthly revenue it does renting shelf space.

So charging a price, any price, creates a mental barrier that most people won't bother crossing. Free, in contrast, speeds right past that decision, increasing the number of people who will try something. What Free grants, in exchange for forsaking direct revenues, is the potential of mass sampling.

After examining mental transaction costs, Clay Shirky, a writer and NYU lecturer, concluded that content creators would be wise to give up on dreams of charging for their offerings:

> For a creator more interested in attention than income, free makes sense. In a regime where most of the participants are charging, freeing your content gives you a competitive advantage. And, as the drunks say, you can't fall off the floor. Anyone offering content free gains an advantage that can't be beaten, only matched, because the competitive answer to free—"I'll pay you to read my weblog!"—is unsupportable over the long haul.
>
> Free content is thus what biologists call an evolutionarily stable strategy. It is a strategy that works well when no one else is using it—it's good to be the only person offering free content. It's also a strategy that continues to work if everyone is using it, because in such an environment, anyone who begins charging for their work will be at a disadvantage. In a world of free content, even the moderate hassle of micropayments greatly damages user preference, and increases their willingness to accept free material as a substitute.

So on a psychological basis (and all economics is rooted in psychology), if there's a way to take the whole "is it worth it?" question off the table, it pays to do so. Note that there are other mental transaction costs to Free—from worrying if it's *really* free to weighing nonmonetary costs such as considering the environmental impact of a free newspaper or just fearing that you'll look like a cheapskate. (One friend tells me the giveaway furniture he puts outside his house is only taken at night.) But those costs aside, taking money out of the equation can greatly increase participation.

Venture capitalist Josh Kopelman of First Round Capital looked at this psychological barrier to paying and realized it made nonsense of the usual teaching about pricing strategy. Rather than supply-and-demand curves turning price into a classic econ 101 calculation, there are really two markets: free and anything else. And the difference between the two is profound. In a sense, what Free does is bend the demand curve. As Wharton professor Kartik Hosanagar says: "The demand you get at a price of *zero* is many times higher than the demand you get at a very low price. Suddenly, the demand shoots up in a nonlinear fashion."

Kopelman called this the "penny gap." Entrepreneurs often come to him, he has said, with business plans that assume they will make their money from subscriptions, and that 5 percent of the people who sample their wares will pay. However, that's rarely the case, as Kopelman explains:

> Most entrepreneurs fall into the trap of assuming that there is a consistent elasticity in price—that is, the lower the price of what you're selling, the higher the demand will be. So you end up with hockey stick looking revenue charts that go up and to the right, all supported by an "it only costs $2 per month" business plan.
>
> The truth is, scaling from $5 to $50 million is not the toughest part of a new venture—it's getting your users to pay you anything at all. The biggest gap in any venture is that between a service that is free and one that costs a penny.

So from the consumer's perspective, there is a huge difference between cheap and free. Give a product away and it can go viral. Charge a single cent for it and you're in an entirely different business, one of clawing and scratching for every customer. The truth is that zero is one market and any other price is another. In many cases, that's the difference between a great market and none at all.

THE COST OF ZERO COST

Traditionally, economics had little to say about Free, since it technically didn't exist in the domain of money at all. But in the 1970s, a new branch of economics emerged that looked at the psychology driving economic behavior. Called "behavioral economics," today the field ranges from game theory to experimental economics. Ultimately, what it tries to explain is why we make the economic choices we do, even when they aren't necessarily the most rational ones.

In *Predictably Irrational,* Dan Ariely describes several experiments he and his colleagues have conducted to try to understand just why this word "free" is so powerful. "Zero is not just another price, it turns out," he writes. "Zero is an emotional hot button—a source of irrational excitement." It's easy to say, but difficult to measure, which is why Ariely set out to do just that.

The first experiment involved chocolate. (Note: Behavioral economists have limited budgets and limited time, so a lot of their experiments involve a folding table, candy, and random college students. So take the results as directionally interesting rather than rigorously quantitative.) The researchers sold two kinds of treats: prized Lindt truffles from Switzerland and ordinary Hershey's Kisses. They priced the Lindt truffles at 15 cents (about half the wholesale price) and the Kisses at 1 cent. The customers behaved pretty rationally, calculating that the difference in quality of the two chocolates more than made up for their difference in price: 73 percent chose the truffle and 27 percent chose the Kiss.

Then Ariely introduced Free into the equation, lowering the price of both chocolates by 1 cent. Now the Lindt truffle was 14 cents and the Kiss was free. Suddenly the humble Kiss became a hit. Sixty-nine percent chose it over the truffle. Nothing about the price/quality calculus had changed—the two chocolates were still priced 14 cents apart. But the introduction of zero caused the customers to reverse their preference.

The psychologically confusing thing in this case is the comparison between two products, one of which is free. Sometimes Free makes perfect sense, as in the case of a bin of free athletic socks in a department store. There's little downside to taking as many as you want (aside from looking like a bit of a miser). But imagine if you went into the store expressly to buy a pair of socks with a nicely padded heel and gold toe. As you reach the sock section, you are distracted by the free version and you end up walking out of the store with something you didn't want (socks with no padding or gold toe) simply because they were free.

What is it about Free that is so enticing? Ariely explains:

> Most transactions have an upside and a downside, but when something is FREE! we forget the downside. FREE! gives us such an emotional charge that we perceive what is being offered as immensely more valuable than it really is. Why? I think it's because humans are intrinsically afraid of loss. The real allure of FREE! is tied to this fear. There's no visible possibility of loss when we choose a FREE! item (it's free). But suppose we choose the item that's *not* free. Uh-oh, now there's a risk of having made a poor decision—the possibility of loss. And so, given the choice, we go for what is free.

There are similar experiments at a larger scale going on every day around us, often by accident. One such example is Amazon's free shipping. As anyone who has used the online retailer knows, you can often get free shipping when the total purchase is more than $25. Amazon's hope is that if you had originally planned to buy a single book for $16.95, the free-shipping offer will entice you to add a second book to your order to bring the total purchase to more than $25. When Amazon rolled this out, it worked great: Sales of second books skyrocketed. Well, everywhere except in France.

What was different about the French? It turns out that they were presented with a slightly different offer. When Amazon rolled out free shipping across all of its national sites, the French one mistakenly set the shipping price to 1 franc, or about 20 cents. That tiny amount completely eliminated the second-book effect. When Amazon fixed this and France joined the other countries in offering free shipping, the French consumers behaved like everyone else and decided to add the second book to their shopping cart.

(Interestingly, Amazon was actually sued for this. A 1981 French law, pushed through by then-minister of culture Jack Lang, forbids booksellers from offering discounts of more than 5 percent off the list price. In 2007, the French Booksellers Union took Amazon to court, arguing that it was exceeding that discount when the free shipping was factored in. The union won, and Amazon was charged $1,500 a day in fines, which, to its credit, it decided to pay rather than eliminate the offer. After all, Free would surely bring in more than enough to make up the difference.)

Zappos, the online shoe retailer, goes even further: It offers free shipping *both* ways: to you and, if you want to return the shoes, back to the warehouse. The point is to eliminate the psychological barrier to buying shoes online, which is that they may not fit. What Zappos wants you to do (really!) is to order several pairs of shoes just to try them on at home. Hopefully, you'll like a pair or two and send the rest back; you only pay for what you keep. The cost of the shipping is built into Zappos's prices, which are not the lowest around, but to its many happy customers, the convenience is worth it.

From a psychological perspective, the use of Free in Zappos's case is simply risk reduction. The only reason to drive to a shoe store is to know that the shoes fit and look good on your feet. By bringing the shoes to you at no additional cost, Zappos reaches risk parity with a physical store, and gains convenience advantage. The only problem, says CEO Tony Hsieh, is that many people still feel guilty about ordering more shoes than they want and sending them back. It wouldn't be a problem if they just didn't send them back (that's a sale!), but rather if they won't order the shoes in the first place, anticipating their guilt when they do send most of them back.

Once again, the enemy of Free is waste. To order shoes you don't really want and send them back feels wasteful, and indeed it is, from the labor of the workers and delivery people involved to the carbon emitted in the transportation. Simply taking money out of the equation isn't enough to fully eliminate the perception of a price, in this case an amorphous social and environmental cost rather than a direct hit to your wallet.

Behavioral economists explain much of our perplexing responses to Free by distinguishing the decisions made in the "social realm" from those made in the "financial realm." Zappos's shipping is free in the financial realm, but not free in the social realm, where our brains try to calculate the net social cost of sending back five pairs of shoes for the one we keep. It's an impossible calculus, and in the face of that, some consumers just shut down: They don't take the offer, even though it's free.

Ariely demonstrated the distinction between these two realms with another experiment: He put six-packs of Coke in college dorm refrigerators. He also left plates of money. People quickly took the Coke, but didn't touch the money. They treated the Coke as "free," even though they know it costs money. But taking actual money felt like stealing.

NO COST, NO COMMITMENT

I was at a conference recently at Google, where they famously offer racks of free snacks, from healthy trail bars to distinctly unhealthy jelly beans. This was a scientific meeting, and it was attended mostly by non-Googlers, largely academics. They kept returning to the rack, understandably impressed by the opulent display of gratis goodies. By the end of the first day, there were half-eaten bags of snacks everywhere.

It's interesting to imagine how that would have been different had Google charged a price for those snacks, even just a dime. I'll bet that a lot less would have been taken, and a lot more people would have finished what they took. Also, I bet they would have been happier with their decision to take a snack. They would have thought about whether they *really* wanted one, and probably waited until they were hungry. And they certainly wouldn't have felt as gross about their hasty decision to eat the snacks (as I did when I absentmindedly grabbed a handful of ginger chews and shoved them into my mouth).

This is one of the negative implications of Free. People often don't care as much about things they don't pay for, and as a result they don't think as much about how they consume them. Free can encourage gluttony, hoarding, thoughtless consumption, waste, guilt, and greed. We take stuff because it's there, not necessarily because we want it. Charging a price, even a very low price, can encourage much more responsible behavior.

The authors of the Penny Closer blog tell the story of a friend who volunteers for a charity that provides people who are down on their luck with transportation—free bus tickets, to be exact. Unfortunately, these tickets, which cost the charity $30 each, are frequently lost. So the charity instituted a new rule—all tickets would cost $1 to help offset the costs of replacement. Suddenly, people lost fewer tickets. Just the act of paying $1 changed how people viewed the ticket. Since they had invested in it, clients seemed to be more careful not to lose it. Even though it was inherently worth something before they had to spend $1 on it, the ticket was somehow worth *even more* now.

The flip side of both these stories is that the imposition of a price, no matter how low, typically decreases participation, often radically. In the Google case, people would take far fewer snacks if they had to pay. In the case of the charity, it distributed far fewer bus tickets. That is the trade-off of Free: Free is the best way to maximize the reach of some product or service, but if that's not what you're ultimately trying to do (Google is not trying to maximize snack food consumption), it can have counterproductive effects. Like every powerful tool, Free must be used carefully lest it cause more harm than good.

THE TIME/MONEY EQUATION

At some point in your life, you may wake up and realize that you have more money than time. You will then realize that you should start doing things differently, which means not walking four blocks to find an ATM that doesn't charge a fee, driving forever to find cheaper gas, or painting your own house.

This same calculus is the foundation of a big part of the freemium economy (see page 26). We often see it in free-to-play online games, such as Maple Story, where you can buy tools like "teleportation stones" to quickly get from one place to another without a long slog or wait for a bus. Most of these paid digital assets don't make you a better player, but they do allow you to become a better player faster.

If you're a kid, you probably have more time than money. That's the force behind MP3 file trading, which is kind of a hassle but is free (albeit illegal!). As Steve Jobs famously pointed out, if you download music from peer-to-peer services, you're likely to deal with problematic file formats, missing album information, and the chance that it's the wrong song or a poor quality version. The time it takes to avoid paying means "you're working for less than minimum wage," he noted. Nevertheless, if you're time-rich and money-poor, that makes sense. Free is the right price for you.

But as you get older, the equation reverses and $0.99 here and there no longer seems as big a deal. You migrate into a paying customer, the premium user in the freemium equation.

One of my side projects is an open source hardware company called DIY Drones (developing and selling aerial robotics technology). You're likely familiar with the concept of open source software, but the new idea of extending that to hardware—from circuit boards all the way up to consumer electronic gadgets like Google's Android phone—is just now emerging.

Even in its nascent form, open source hardware is a really interesting example of how to make money from Free. It adds a new dimension to the open source software world, because it's about atoms (which have real marginal costs), not just bits of information that can be propagated at nearly no expense.

The way most open source hardware companies work is this: All the plans, printed circuit board files, software, and instructions are free and available to all. If you want to build your own (or, even better, improve on a design), you're encouraged to do so. But if you don't want the hassle or risk of doing it yourself, you can buy a premade version that's guaranteed to work.

For instance, take the Arduino open source microprocessor that DIY Drones's autopilots are based on. You can build your own, with full instructions, which can be found at arduino.cc. Or you can buy one. Most people do the latter. The Arduino team make their money from a certification license fee they charge the companies and retailers that make and sell the boards.

You can build a good business on this model, as Limor Fried has shown with Adafruit Industries, her electronics kit retail/design company. She and her business partner, Phillip Torrone, have a simple business model built around Free, which I have shamelessly copied for DIY Drones.

Here's how it works:

1. Build a community around free information and advice on a particular topic.
2. With that community's help, design some products that people want, and return the favor by making the products free in raw form.
3. Let those with more money than time/skill/risk-tolerance buy the more polished version of those products. (That may turn out to be almost everyone.)
4. Do it again and again, building a 40 percent profit margin into the products to pay the bills.

It's really just as simple as that. As Torrone says, "I can't imagine doing a book, a video, or a magazine unless I had a community that would rally along the way. In the end it always seemed to be about a story—people like to see the beginning, middle, end, and plot of something—and if there's a buy button somewhere, they sometimes click it and reward us for working hard."

When you think about this, it's another example of the psychology of Free, in two ways. The first is the mental calculation we do when we value our time. Remember Steve Jobs's assertion that you're not even paying yourself minimum wage if you choose to take the time to wade through all the messy metadata that comes with file trading? Jobs was saying that the case for paying $0.99 for a song is that it's a time saver (aside from all the other arguments about legality and fairness).

The second reason you might want to pay for something is to lower the risk of it not being what you want. Prices come with guarantees, while Free typically doesn't. In the case of Adafruit, that's what they're selling with their premade electronics. You can be sure that they'll work, which is not the case if you're soldering it together yourself.

But Free can help instill confidence, too. Again, let's take Adafruit. The fact that there is a free version and an open source version of the product available means that you can inspect it and try it without risk. Plus, you know that you can modify it if it doesn't precisely fit your needs. Also, the fact that there is a free version has attracted a larger community of users. The knowledge that so many others have been drawn to the product and are there to help if you have problems is re-assuring, too. (In psychology, this is called "mimetic desire," which ba-sically says that we want to do things that other people do because their decisions validate our own, which explains everything from herd behavior to hipster trucker hats.)

This is why Free works so well in conjunction with Paid. It can ac-commodate the varying psychologies of a range of consumers, from those who have more time than money to those who have more money than time. It can work for those who are confident in their skills and want to do it themselves, and for those who aren't and want someone to do it for them. Free plus Paid can span the full psychology of con-sumerism.

THE PIRATE BRAIN

A final form of Free that we haven't yet talked about in detail is piracy. Piracy is a special form of theft, one that is often considered by pirates and consumers of pirated goods alike to be a relatively victimless crime. (I won't attempt to discuss here whether I think they're right or not; we'll just look at how they see it, from a psychological perspective.) The argument is that a pirated good rarely substitutes for the authentic original. Instead, it allows the product to reach populations that can't afford the original or otherwise wouldn't have bought it.

The reason piracy is a special class of theft is that the costs to the rightful owner are intangible. If you make a music album that is then pirated, the pirates haven't taken something you own, they have *reproduced* something you own. This is an important distinction, which boils down to the reality that you don't suffer a loss but rather a *lesser gain*. The costs are, at most, the opportunity costs of sales not made because the original was competing with pirated versions in the marketplace. (We'll discuss this further in Chapter 14, which looks at pirate markets in China, where you'll see that the results are not always entirely negative for the rightful owner.)

Piracy is a form of imposed Free. You may not have intended your product to be free, but the marketplace thrust Free upon you. For the music industry and much of the software industry, this is an everyday reality. Free has become the de facto price regardless of every effort to stop it.

One software developer decided to find out why. Cliff Harris creates video games priced at what he thought was a very reasonable $20. Yet his games were being pirated constantly. Why?

He asked the readers of Slashdot, a popular technology discussion site. He got hundreds and hundreds of replies, few of them shorter than one hundred words. "It was," he said, "as if a lot of people have waited a long time to tell a game developer the answer to this question."

Kevin Kelly reported on the experiment:

He found patterns in the replies that surprised him. Chief among them was the common feeling that his games (and games in general) were overpriced for what buyers got—even at $20. Secondly, anything that made purchasing and starting to play difficult—copy protection, digital rights management (DRM), or complicated online purchasing routines—anything at all standing between the impulse to play and playing in the game itself was seen as a legitimate signal to take the free route. Harris also noted that ideological reasons (rants against capitalism, intellectual property, and "the man," or simply liking being an outlaw) were a decided minority.

Much to his credit, the sincere responses to his question changed Harris's mind. He decided to alter his business model. He reduced the price of his games in half (to $10). He removed the little copy protection he had been using. He promised to make his Web store easier to use, maybe even with one-click checkout. He decided to increase the length of his free demos. Most importantly, he had the revelation that he needed to increase the quality of his games.

In a sense, the people in the marketplace were telling him that they valued his games at less than he thought they were worth. He realized any efforts to fight this would be fruitless unless people thought the games were worth more.

The lesson from Harris's experience is that in a digital marketplace, Free is almost always a choice. If you don't offer it explicitly, others will typically find a way to introduce it themselves. When the marginal cost of reproduction is zero, the barriers to Free are mostly psychological—fear of breaking the law, a sense of fairness, an individual's calculation on the value of his or her time, perhaps a habit of paying or ignorance that a free version can be obtained. Sooner or later, most producers in the digital realm will find themselves competing with Free. Harris understood that and figured out how to do it better. With his survey, he looked into the mind of the pirate and saw a paying customer looking for a reason to come out.

5

TOO CHEAP TO MATTER

The Web's Lesson: When Something Halves in Price Each Year, Zero Is Inevitable

IN 1954, at the dawn of nuclear power, Lewis Strauss, the head of the Atomic Energy Commission, stood before a group of science writers in New York City and foretold great things to come. Diseases would be conquered and we would come to understand what causes man to age. People would soon travel "effortlessly" over the seas and through the air at great speeds. Great periodic regional famines would become a matter of history. And, most famously, he predicted, "It is not too much to expect that our children will enjoy in their homes electrical energy too cheap to meter."

These were optimistic times: It was the beginning of the space age, modern medicine was conquering ancient afflictions, chemistry was bringing "better living" and feeding the planet, and the Information Age was dawning with infinite possibility. Anything that could be invented would be invented and then quickly branded, packaged, and sold to an emerging class of free-spending consumers.

The postwar optimism that science and technology could launch a prosperous era of unprecedented growth extended from national pride to domestic bliss. The power of human thought and clever machinery promised to liberate us from household drudgery and end war. The question wasn't whether we'd live in space colonies but what we'd wear there. The Jetsons were a joke, but no more so than the Flintstones; the notion we'd someday have space taxis and robot butlers was as certain as the fact we'd once dwelled in caves.

And indeed, the postwar science and technology boom did set us on a path of increasing productivity and economic growth at a rate never before seen. But it wasn't quite as rosy as Strauss predicted. Electricity didn't get too cheap to meter.

Although the fuel costs of uranium were low compared to coal, the initial costs of building the reactors and power plants turned out to be much higher. Waste disposal was and remains an unsolved problem. And an expensive and risky proposition became doubly so after Three Mile Island and Chernobyl.

Today, nuclear energy costs about the same as coal, which is to say that it didn't change the economics of electricity one bit.*

But what if Strauss had been right? What if electricity had, in fact,

---

*Like all famous quotes, Strauss's is often misunderstood. First, he was probably talking about hydrogen fusion, not uranium fission. Then, as today, fusion was decades away from being viable. Fission (what's known as "nuclear power"), on the other hand, was already in the works and everyone, including Strauss, knew that it probably would be more expensive than coal, given the high capital costs of setting up the plants.

Second, "Too cheap to meter" doesn't mean free: It just means too cheap to monitor closely. Indeed, some buildings built around that time, including the World Trade Center, were designed without light switches in each office. Instead, building managers could just turn on and off whole floors, like a Christmas tree. Electricity was expected to be too cheap to bother thinking about.

As an aside, Strauss was a controversial character for more than his flair for hyperbole. He was a strong proponent of the hydrogen bomb, which put him in conflict with Robert Oppenheimer, the regretful father of the atomic bomb. He famously testified against Oppenheimer in a congressional witch hunt that led to Oppenheimer losing his security clearance. Strauss told President Eisenhower that he would only accept the position of AEC chair if Oppenheimer played no role in advising the agency. He explained that he didn't trust Oppenheimer partly because of the scientist's consistent opposition to the superbomb. Within days of being sworn into office in July 1953, Strauss had all classified AEC material removed from Oppenheimer's office.

But he got his comeuppance: According to his bio, "Over the years Strauss' arrogance and his insistence that he was always right made him unpopular on Capitol Hill. In 1959, after two months of exhausting hearings, the Senate rejected his nomination to be Secretary of Commerce. The ordeal was publicly humiliating for Strauss, especially after he was caught lying under oath."

become virtually free? The answer is that everything that electricity touched—which is to say nearly everything—would have been transformed. Rather than balance electricity against other energy sources, we'd now use electricity as much as we could—we'd waste it, because it would be so cheap that it wouldn't be worth worrying about efficiency.

All buildings would be electrically heated, never mind the thermal conversion rate. We'd all be driving electric cars. (Free electricity would be incentive enough to develop the efficient battery technology to store it.) Massive desalination plants would turn seawater into all the freshwater anyone could want, allowing us to irrigate vast inland swaths and turn deserts into fertile acres.

Because two of the three major inputs to agriculture—air and sun—are free, and water would now join them, we could begin to grow crops far in surplus to our food requirements, and many of them would be the feedstocks for biofuels. In comparison, fossil fuels would be seen as ludicrously expensive and dirty. So net carbon emissions would plummet. (Plants take carbon out of the atmosphere before they release it again in burning, while oil and coal add more carbon.) The phrase "global warming" might never have entered the language.

In short, "too cheap to meter" would have changed the world.

Unlikely? For electricity, perhaps (although who knows what solar energy may someday bring?). But today there are three other technologies that touch nearly as much of our economy as electricity does: computer processing power, digital storage, and bandwidth. And all three really are getting too cheap to meter.

The rate at which this is happening is mind-boggling, even nearly a half century after Gordon Moore first spotted the trend line now called Moore's Law. Even more astounding, processing power—the one Moore tracked—is actually improving at the slowest pace of the three. Semiconductor chips roughly double the number of transistors they can hold every eighteen months. (That's why for the same price every two years or so you can buy an iPod that holds twice as much music as the last one.) Hard drive storage is getting better even faster: The number of bytes that can be saved on a given area of a hard disk doubles about every year, which is why you can now store hundreds of hours of video on your TiVo. But the fastest of all three is bandwidth: The speed at which data can be transferred over a fiber-optic cable doubles every nine months. That's why you don't even need TiVo anymore—you can watch all the TV you want, when you want it, with streaming online video services such as Hulu.

For each of these technologies there is an economic corollary that is, if anything, even more powerful: Costs halve at the same rate that capacity, speed, etc., doubles. So that means that if computing power for a given price doubles every two years, a given unit of computing power will fall in price by 50 percent over the same period.

Take the transistor. In 1961, a single transistor cost $10. Two years later, it was $5. Another two years later, when Moore published his prediction in the April 1965 issue of *Electronics* magazine, it was $2.50. By 1968, the transistor had fallen to $1. Seven years later, it was 10 cents. Another seven years and it was a penny, and so on.

Today, Intel's latest processor chips have about 2 billion transistors and cost around $300. So that means each transistor costs approximately 0.000015 cents. Which is to say, too cheap to meter.

This "triple play" of faster, better, cheaper technologies—processing, storage, and bandwidth—all come together online, which is why today you can have free services like YouTube—essentially unlimited amounts of video that you can watch without delay and with increasingly high resolution—that would have been ruinously expensive just a few years ago.

Never in the course of human history have the primary inputs to an industrial economy fallen in price so fast and for so long. This is the engine behind the new Free, the one that goes beyond a marketing gimmick or a cross-subsidy. In a world where prices always seem to go up, the cost of anything built on these three technologies will *always* go down. And keep going down, until it is as close to zero as possible.

ANTICIPATE THE CHEAP

When the cost of the thing you're making falls this regularly, for this long, you can try pricing schemes that would seem otherwise insane. Rather than sell it for what it costs today, you can sell it for what it will cost *tomorrow*. The increased demand this lower price will stimulate will accelerate the curve, ensuring that the product will cost *even less than expected* when tomorrow comes. So you make more money.

For instance, in the early 1960s, Fairchild Semiconductor was selling an early transistor, called the 1211, to the military. Each transistor cost $100 to make. Fairchild wanted to sell the transistor to RCA for use in their new UHF television tuner. At the time RCA was using traditional vacuum tubes, which cost only $1.05 each.

Fairchild's founders, the legendary Robert Noyce and Jerry Sanders, knew that as their production volume increased, the cost of the transistor would quickly go down. But to make their first commercial sale they needed to get the price down immediately, before they had any volume at all. So they rounded down. Way down. They cut the price of the 1211 to $1.05, right from the start, before they even knew how to make it so cheaply. "We were going to make the chips in a factory we hadn't built, using a process we hadn't yet developed, but the bottom line was: We were out there the next week quoting $1.05," Sanders later recalled. "We were selling into the future."

It worked. By getting way ahead of the price decline curve, they made their goal of $1.05 and took 90 percent of the UHF tuner market share. Two years later they were able to cut the price of the 1211 to 50 cents, and still make a profit. Kevin Kelly, who described this effect in his book *New Rules for the New Economy*, calls this "anticipating the cheap."

Imagine if Henry Ford had enjoyed the same trend in his Model T factory. It seems almost impossible: How could physical stuff such as a car fall in price the way digital technology does? Every year, we'd have to get twice as good at extracting ore from the ground and turning it into metals. All the components that go into a car would have to get cheaper, like semiconductor chips, obeying some sort of Moore's Law of windshield wipers and transmission machining. Workers would have to agree to cut their salaries in half each year, or half of them would have to be replaced by robots.

But if you had been alive in the first few decades of the automobile industry, this wouldn't be impossible to imagine at all. Between 1906 and 1918, automobile "quality-adjusted" prices (the performance of the car per dollar) fell by about 50 percent every two years, so by the end of that period, an equivalent car cost just one-fifth what it had a decade earlier.

By moving from handcrafting to an assembly line powered by electric motors, Ford was able to lower the cost of muscle power. Then, by switching from custom-crafted parts to standard manufactured components, he lowered the cost of labor again and sold millions of mass-produced cars.

But that remarkable cost-decline curve, a product of Henry Ford's groundbreaking production line techniques, couldn't last. The price/performance improvements of cars slowed, and today they amount to just a few percent a year. We have indeed become much better at extracting ore from the ground, and half of automotive workers really have been replaced by robots, but it didn't happen overnight. Cars do get cheaper and better, but at nothing like the pace of digital technology. Today, a car is still an expensive item.

From an environmental perspective, that's no bad thing. Even if it were possible for physical goods to fall in price as quickly as microchips, the "negative externalities" of the resulting overproduction of stuff would soon be all too apparent. If you've seen Pixar's *WALL-E*, where humans are driven from the planet by the overflowing mountains of trash, you can imagine the problem.

But in the digital realm, where what's created in abundance is ultimately ephemeral bits of information—electrons, photons, and magnetic flux—there's nothing stopping such remarkable doubling laws from playing out to their full effect. And the consequence is, as Moore

HOW CAN A CAR BE FREE?

Just as Ryanair redefined the airline business to be less about selling seats and more about selling travel, Better Place is redefining the auto industry. In an era of high gas prices, people are realizing the cost of a car goes beyond the purchase price—it's also the expense of running it, which can total over $3,000 a year. Taking a page from the mobile phone business, Better Place plans to give away the car, while selling the miles for less than you'd pay with a traditional car.

This model works if gas gets more expensive faster than electricity.

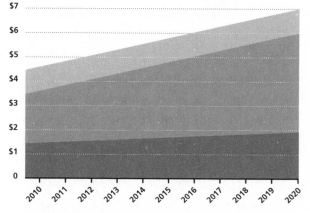

PROJECTED PRICE OF A GALLON OF GAS IN TARGET MARKET

BETTER PLACE PER MILE CHARGE

AVERAGE PRICE OF EQUIVALENT AMOUNT OF ELECTRICITY

Better Place can do this since its cars are electric, and electricity is cheaper than gas. If you sign a three-year contract (and live where the service is available, currently planned for Israel, Denmark, Australia, and the San Francisco Bay Area), Better Place will lease you a car for free. You'll charge it at home and work with a special charging station, and at public spots the car's onboard GPS directs you to. If you're in a rush, you'll be guided to a station where attendants will switch power packs faster than it would take to fill up with gas.

The current gap between gas prices and their per-mile equivalent of electricity in Israel, Better Place's first market, is about $3 per gallon. In countries with high gas taxes and lots of renewable energy, like most of Europe, this gap can be as high as $4. Better Place uses the difference to subsidize its cars.

It foresees two sets of customers: those who buy the car and get the battery for free, and those who get both for free. Better Place wants the former to feel a savings, so it sets its per-mile prices at less than the equivalent gas cost. It's betting gas will

get more expensive faster than electricity, since oil capacity is limited while renewable electricity sources are not. If a driver travels 10,000 miles per year at a cost of $0.15 a mile and Better Place offers $0.12, when the electricity actually costs $0.02, it makes a dime per mile. That's $1,000 in gross annual profit, which pays off the cost of the battery over time (packs last at least a decade). As the price gap between gas and electricity widens, Better Place will return the cost of the batteries faster and make more money.

For the second customer, Better Place will charge more per mile, as much as $0.50 ($15,000 per year), which is enough to make up the cost of car and battery. And the economics only improve as gas gets more expensive compared to electricity.

The non-economic benefits are greater: no greenhouse gas emissions from the car and less dependency on foreign oil. This qualifies Better Place for government/corporate subsidies that will pay for much of the public charging infrastructure where it's first launching. It will expand to regions where the economics make the most sense.

himself pointed out, amazing: "Moore's law is a violation of Murphy's law. Everything gets better and better."

WHY MOORE'S LAW WORKS

Most industrial processes get better over time and scale through an effect known as the learning curve. It's just that those processes based on semiconductors do so much faster and longer.

The term "learning curve" was introduced by the nineteenth-century German psychologist Hermann Ebbinghaus to describe improvements he observed when people memorized tasks over many repetitions. But it soon took on broader meaning. The principle states that the more times a task has been performed, the less time will be required for each subsequent iteration. This relationship was first quantified in 1936 at Wright-Patterson Air Force Base, where managers recorded that every time total aircraft production doubled, the required labor time decreased by 10 to 15 percent.

In the late 1960s, the Boston Consulting Group (BCG) started looking at technology industries and saw improvements that were often faster than simple learning curves could explain. Where the learning curve was mostly about human learning, these larger effects seemed to have more to do with scale: As products were manufactured in larger numbers, the costs fell by a constant and predictable percentage (10 to 25 percent) with every doubling of volume. BCG called this the "experience curve" to encompass institutional learning, ranging from administrative efficiencies to supply chain optimization, as well as the individual learning of the workers.

But starting in the 1970s, price declines in the new field of semiconductors seemed to be happening even faster than the experience curve alone could explain. The original transistors fell at the high end of the BCG rate and kept on falling. During one decade-long period, the Fairchild 1211 transistor's sales increased four thousandfold. That's twelve doublings, which experience-curve theory predicts would lead to a price decline of one-thirtieth the original figure. In fact, the price fell to *one-one-thousandth* that number. There was clearly something more going on.

What's different about semiconductors is a characteristic of many high-tech products: They have a very high ratio of brains to brawn. In economic terms, their inputs are mostly intellectual rather than material. After all, microchips are just sand (silicon) very cleverly put together. As George Gilder, the author of *Microcosm*, puts it:

> When matter plays so small a part in production, there is less material resistance to increased volume. Semiconductors represent the overthrow of matter in the economy.

In other words, ideas can propagate virtually without limit and without cost. This, of course, is not new. Indeed, it was Thomas Jefferson, father of the patent system (and a lot more), who put it better than anyone:

> He who receives an idea from me, receives instruction himself without lessening mine; as he who lights his taper at mine, receives light without darkening me.

The point: Ideas are the ultimate abundance commodity, which propagates at zero marginal cost. Once created, ideas want to spread far and wide, enriching everything they touch. (In society, such spreading ideas are called "memes.")

But in business, companies make their money by creating an artificial scarcity in ideas through intellectual property law. That's what patents, copyright, and trade secrets are: efforts to hold back the natural flow of ideas into the population at large long enough to make a profit. They were created to give inventors an economic incentive to create, a license to charge monopoly rent for a limited time, so they can get a return on the work they put into the idea. But ultimately, patents expire and secrets get out; ideas cannot be held back forever.

And the more products are made of ideas, rather than stuff, the faster they can get cheap. This is the root of the abundance that leads to Free in the digital world, which we today shorthand as Moore's Law.

However, this is not limited to digital products. Any industry where information becomes the main ingredient will tend to follow this compound learning curve and accelerate in performance while it drops in price. Take medicine, which is shifting from "we don't know why it works, it just does" (there's a reason it's called drug "discovery") to a process that starts with the first principles of molecular biology ("now we know why it works"). The underlying science is information, while observed efficacy is just anecdote. Once you understand the basics, you can create an abundance of better drugs, faster.

DNA sequencing is falling in price by 50 percent every 1.9 years, and soon our individual genetic makeup will be another information industry. More and more medical and diagnostic services will be provided by software (which get cheaper, to the point of being free) as opposed to doctors (who get more expensive).

Likewise for nanotechnology, which promises to turn manufacturing into yet another information industry, as custom-designed molecules self-assemble. As energy shifts from burning fossil fuels to using photovoltaic cells to convert sun into electricity on a utility scale, or designing enzymes that can convert grass into ethanol, it will be an information industry, too. In each case, industries that have nothing to do with computer processing start to show Moore's Law–like exponential growth (and price declines) once they, too, become more brains than brawn.

MEAD'S LAW

As it happens, Gordon Moore did not coin the law named after him. The man who did, Caltech professor Carver Mead, was the first to focus on the economic corollary to Moore's doubling rule for transistor density: If the amount of computer power for a given cost doubles every two years, then the cost of a given unit of computing power must halve over the same period. More importantly, he was the first to really consider what this meant for how we thought about and used digital technology. And he realized that we were thinking about it all wrong.

In the late 1970s, Mead was teaching semiconductor design at Caltech, defining the principles of integrated circuits that would become known as Very Large Scale Integration (VLSI), which pretty much defined the world of computing that we have today. Like Moore before him, he could see that the eighteen-month doublings in performance would continue to stretch out as far as anyone could see. This was driven not just by the standard learning and experience curves, but also by what he called the "compound learning curve," which is the combination of learning curves and frequent new inventions.

For more than a half century, semiconductor researchers have come up with a major innovation every decade or so that kicks the industry

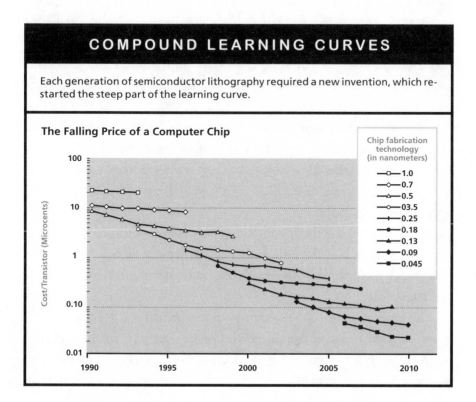

COMPOUND LEARNING CURVES

Each generation of semiconductor lithography required a new invention, which re-started the steep part of the learning curve.

The Falling Price of a Computer Chip

Chip fabrication technology (in nanometers)
- 1.0
- 0.7
- 0.5
- 03.5
- 0.25
- 0.18
- 0.13
- 0.09
- 0.045

into the sharp-decline part of the curve again. As one production process nears the tail of its efficiency improvement cycle, the incentive to come up with something radically new and better increases. And because there is, as physicist Richard Feynman said, "a lot of room at the bottom" of the atomic-scale world that opened up with the new physics of the late twentieth century, researchers have been able to find these new ways with almost spooky regularity.

Each time, whether it's a new material, a new etching process, a new chip architecture, or an entirely new dimension such as parallel processing, the learning curve starts again at its most vertiginous slope. When you combine all these innovations and learning curves across the entire computer industry, you end up with a pace of cost decline never before seen. Transistors, like almost any other unit of computing capacity you can pick, march inevitably toward a price of zero.

What Mead realized is that this economic effect carried with it a moral imperative. If transistors are becoming too cheap to meter, then we should stop metering them and otherwise cease thinking about their cost. We should switch from conserving them on the assumption that they are a scarce commodity to treating them like the abundant commodity they are. In other words, we should literally start "wasting" them.

"Waste" is a dirty word, and that was especially true in the IT world of the 1970s. An entire generation of computer professionals had been taught that their job was to dole out expensive computer resources sparingly. In the glass-walled facilities of the mainframe era, these systems operators exercised their power by choosing whose programs should be allowed to run on the costly computing machines. Their role was to conserve transistors, and they not only decided what was worthy but also encouraged programmers to make the most economical use of their computer time.

This priesthood—the sysadmins—ruled the early information age. If you wanted to use a computer, you had to get past them. And that meant writing a program that conformed to their standards of what was an appropriate use of IT resources. Software should be focused on business objectives, efficient in its use of CPU cycles, and modest in its ambitions. If you passed that test, they might accept your punch cards through the slot in the door and, two days later, return to you a print-out of your error messages so you could start the process again.

As a result, early developers focused their code on running their core algorithms efficiently and gave little thought to user interface. This was the era of the command line, and the job of software was to serve the central processing unit, not the other way around.

Engineers of the time understood Moore's Law on one level: They knew it would bring computers that were smaller and cheaper than the mainframes of the day. Indeed, it was not too much to imagine comput-ers becoming so small and cheap that a regular family could have one in their home. But why would anyone want that? After much pondering, the computing establishment of the late sixties could think of only one reason: to organize recipes. The world's first personal computer, a stylish kitchen appliance offered by Honeywell in 1969, did just that—and it even came with integrated counter space. The Honeywell was featured in that year's Neiman Marcus catalog, selling at the bargain price of $10,600 despite the fact that the only input method was toggle switches on the front panel and the housewife would have to speak hexadecimal. It's unclear whether anyone ever bought one.

And here was Mead, telling programmers to embrace waste. They scratched their heads—how do you waste computer power?

THE MOUSE THAT ROARED

It took Alan Kay, an engineer working at Xerox's Palo Alto Research Center in the 1970s, to show them the way. Rather than conserve tran-sistors for core-processing functions, he developed a computer concept— the Dynabook—that would frivolously deploy silicon to do playful things on the screen: draw icons, steer pointers with a mouse, divide a screen into windows, and even add animations for no function other than to look cool.

The purpose of this profligate eye candy? To make computers easier to use for regular folks, including children. Kay's work on the graphical user interface (GUI) became the inspiration for the Xerox Alto and then the Apple Macintosh, which changed the world by opening computing to the rest of us.

What Kay realized was that a technologist's job is not to figure out what technology is good for. Instead it is to make technology so cheap, easy to use, and ubiquitous that anybody can use it, so that it propagates around the world and into every possible niche. We, the users, will figure out what to do with it, because each of us is different: different needs, different ideas, different knowledge, and different ways of interacting with the world.

Kay, by showing the way to democratize computing, made it possible to take Moore's phenomenon out of the glass box and into every home, car, and pocket. This collective exercise in exploring the potential space of computing has brought us everything from digital photography to video games, from TiVos to iPods. (Tellingly, organizing recipes is not high on many people's lists.)

The engineers brought us the technical infrastructure of the Internet and Web—TCP/IP and http://—but *we* were the ones who figured out what to do with it. Because the technology was free and open to all, we, the users, experimented with it and together we populated it with our content, our ideas, and ourselves. The technologists invented the pot, but we filled it.

Of course, cheap technology is not free technology. Powerful computers were expensive in Kay's day, and they remain expensive today, as the poor CIO who just shelled out six figures to buy another rack of servers will be the first to tell you. Technology sure doesn't feel free when you're buying it by the gross. Yet if you look at it from the other side of the fat pipe, the economics change. That expensive bank of hard drives (high fixed costs) can serve tens of thousands of users (low marginal costs).

Today's Web is all about scale, finding ways to attract the most users for centralized resources, spreading those costs over larger and larger audiences as the technology gets more and more capable. It's not about the cost of the equipment in the racks at the data center; it's about what that equipment can do. And every year, like some sort of magic clockwork, it does more and more for less and less, bringing the marginal costs of technology in the units that we individuals consume closer to zero.

What Mead and Kay anticipated had a profound effect on computation-based industries. It meant software writers, liberated from worrying about scarce computational resources like memory and CPU cycles, could become more and more ambitious, focusing on higher-order functions such as user interfaces and new markets such as entertainment. The result was software of broader appeal, which brought in more users, who in turn found even more uses for computers. Thanks to that wasteful throwing of transistors against the wall, the world was changed.

What's interesting is that transistors (or storage, or bandwidth) don't have to be completely free to invoke this effect. At a certain point, they're cheap enough to be safely disregarded. The Greek philosopher Zeno wrestled with this concept in a slightly different context. In Zeno's dichotomy paradox, you run toward a wall. As you run, you halve the distance to the wall, then halve it again, and so on. But if you continue to subdivide space forever, how can you ever actually reach the wall? (The answer is that you can't: Once you're within a few nanometers, atomic repulsion forces become too strong for you to get any closer. As for the apparent mathematical paradox, Newton solved that with the invention of integral calculus.)

In economics, the parallel is this: If the unitary cost of technology ("per megabyte" or "per megabit per second" or "per thousand floating-point operations per second") is halving every eighteen months, when does it come close enough to zero to say that you've arrived and can safely round down to nothing? The answer: almost always sooner than you think.

What Mead understood was that a psychological switch should flip as things head toward zero. Even though they may never become entirely free, as the price drops there is great advantage to be had in treating them as if they *were* free. Not too cheap to *meter*, as Strauss foretold, but too cheap to *matter*.

IRON AND GLASS

The story of the semiconductor has largely become the fable for the digital economy, but as I noted above, the truth is that two related technologies—storage and bandwidth—have outpaced it in the race to the bottom.

The first, digital storage, is based not on etching silicon into finer and finer lines, but instead on getting magnetic particles on a metal platter to lie one way or another. This is the way the hard drive in your personal computer works: A tiny electromagnet floats a few atoms' width above a spinning disk and traces out spirals on that disk, flipping the magnetic particles underneath to represent 1s or 0s (put enough of them together and you've got that PowerPoint you've been working on or video you just downloaded). The way to pack more bits onto a platter is to make those tracks smaller, which is done by having a tinier, higher-power head floating even closer to a disk made up of even smaller, more highly magnetized particles.

This is largely a matter of mechanical assemblies with a precision that puts a Swiss watch to shame, along with a platter made of ferrous (iron) materials that can hold intense magnetic fields. Even though storage is based on different physics than semiconductors, Mead's compound learning curves hold sway. It really has very little to do with the semiconductor effects that Moore was observing, and yet data storage capacity is increasing (and the costs falling) even faster than Moore's Law. Once again the ratio of thought to stuff is high, and the innovations frequent.

The second, bandwidth, taps yet another domain of physics and materials science. Sending data long distances is mostly a matter of photons, not electrons. Optical switches convert the on/off bits of binary code into pulses of laser light at different frequencies, and those "lambdas" (the Greek letter used to denote wavelength) travel in threads of glass so pure that the light bounces off the internal walls for hundreds of miles without loss or distortion.

Here the science is optics, not materials science or mechanical precision. Yet the ratio of intellectual ingredients to physical ones is once again high, so the innovations continue to come frequently to restart the improvement cycle. Again following Mead's compound learning curves, fiber-optic networks and optical switching are improving even faster than processing and storage, with an estimated doubling in price/performance every year.

WHAT ABUNDANCE CAN DO

Bandwidth that's too cheap to meter brought us YouTube, which is quickly revolutionizing (some say destroying) the traditional television industry. Storage that's too cheap to meter brought us Gmail and its infinite inbox, to say nothing of TiVo, Flickr, MySpace, and the iPod.

Before the iPod, nobody was asking to carry an entire music collection in a pocket. But engineers at Apple understood the economics of storage abundance. They could see that disk drives were gaining capacity for the same price even faster than computer processors. Demand for storing massive catalogs of music wasn't driving this—physics and engineering were. But the Apple engineers "listened to the technology," to use Mead's phrase.

They paid particular attention to a 2000 announcement by Toshiba that it would soon be able to make a 1.8-inch hard disk that could store five gigabytes. How much storage capacity is that? Well, if you do the math, that's enough to store a thousand songs on a drive smaller than a deck of playing cards. So Apple simply did what the technology enabled and released that product. Supply created its own demand—consumers may not have thought about carrying their entire music libraries around with them, but when offered the opportunity to do so, the advantages became immediately obvious. Why predict what you're going to want to listen to and upload just that, when you can have it all?

Now that this triple play of technologies—processing, storage, and bandwidth—has combined to form the Web, the abundances have been compounded. One of the dot-com jokes from the late-nineties bubble was that there are only two numbers on the Internet: infinity and zero. The first, at least as it applied to stock market valuations, proved false. But the second is alive and well. The Web has become the land of the free, not because of ideology but because of economics. Price has fallen to the marginal cost, and the marginal cost of everything online is close enough to zero that it pays to round down.

Just as the computer industry took decades to understand the implications of Moore's observation, it will take decades more to understand the compounded consequences of the Internet's connecting processing to bandwidth and storage, the two other horsemen of the zerosphere.

When Lewis Strauss predicted electricity would become too cheap to meter, it already touched every part of the economy. It was mind-blowing to imagine what such abundance might bring. Now information touches nearly as much of the economy as electricity.

Information is how money flows; aside from the cash in your wallet, that's what money is—just bits. Information is how we communicate, as every call is turned into data the moment the words leave our lips. It's the TV and movies we watch and the music we listen to—born digital, and thus transforming like everything else in the world of bits, changing how it's made and how we consume it. Even electricity itself is increasingly becoming an information industry, both in the dispatching core of the grid and at its edge, as "smart grids" turn one-way networks interactive, soon to be regulating demand and both sending and receiving electrons from solar panels and electric cars.

Everything that bits touch is also touched by their unique economic properties—cheaper, better, faster. Make a burglar alarm digital, and now it's just another sensor and communications node on the Internet, with abundant storage, bandwidth, and processing added essentially for free. This is why there is such an incentive to turn things digital: They can suddenly be part of something bigger, something not just operating faster, but accelerating.

Bits are industrial steroids in the same way that electricity was—they make everything cost less and do more. The difference is that they keep working their improvement magic year after year. Not a one-time transformation like electricity, but a continuing revolution, with each new generation of half-the-price, twice-the-performance opening up entirely new possibilities.

But what about that first lesson from economics class, that price is set by supply and demand, not science? Have no fear—that still holds. Supply and demand determine the price for any of these commodities at any given moment. But the long-term pricing trends are determined by the technology itself—the more there is of a commodity, the cheaper it will be. Say's Law (named after the early-nineteenth-century French economist Jean-Baptiste Say) states that "supply creates its own demand," which is another way of saying that if you make a million times as many transistors, the world will find a use for them.

At any given time, the world may want slightly more or slightly less than is currently being produced, and the instantaneous price will reflect that, rising or falling with supply and demand. But in the long term, falling costs of production ensure that the overall trend is down, with momentary supply/demand imbalances just introducing ripples in a line that's inevitably heading toward zero.

So today an entire economy has been built on compound learning curves. It's an astounding thing, one that has taken a generation to understand and will take generations more to fully exploit. But the first recognition of its implications came not from economists but rather from the radical underground of . . . model train hobbyists.

6

"INFORMATION WANTS TO BE FREE"

The History of a Phrase That Defined the Digital Age

IN 1984, journalist Steven Levy published *Hackers: Heroes of the Computer Revolution,* which chronicled the scruffy subculture that had not only created the personal computer (and eventually the Internet) but also the unique social ethos that came with it. He listed seven principles of the "hacker ethic":

1. Access to computers—and anything that might teach you something about the way the world works—should be unlimited and total.
2. Always yield to the Hands-on Imperative!
3. All information should be free.
4. Mistrust authority—promote decentralization.
5. Hackers should be judged by their hacking, not bogus criteria such as degrees, age, race, or position.
6. You can create art and beauty on a computer.
7. Computers can change your life for the better.

Number three, which dates back to 1959, is originally credited to Peter Samson of MIT's Tech Model Railroad Club. The TMRC was the ultimate proto-geek community and perhaps the nerdiest group of humans who had ever assembled to date. Its Wikipedia entry explains why they mattered:

> The club was composed of two groups: those who were interested in the modeling and landscaping, and those who comprised the Signals and Power Subcommittee and created the circuits that made the trains run. The latter would be among the ones who popularized the term "hacker" among many other slang terms, and who eventually moved on to computers and programming. They were initially drawn to the IBM 704, the multimillion-dollar mainframe that was operated at Building 26, but access and time to the mainframe was restricted to more important people. The group really began being involved with computers when Jack Dennis, a former member, introduced them to the TX-0, a three-million-dollar computer on long-term-loan from Lincoln Laboratory. They would usually stake out the place where the TX-0 was housed until late in the night in hopes that someone who had signed up for computer time did not show up.

Levy's book wound up on the radar of Kevin Kelly, who would later become the executive editor of *Wired* magazine (and who remains our "Senior Maverick" and advisor) and Stewart Brand, one-time Merry Prankster and creator of the *Whole Earth Catalog*, perhaps the most influential publication birthed by the counterculture, which was edited by Kelly. In 1983, Brand received an advance of $1.3 million to establish a *Whole Earth Software Catalog*. The idea was for the book to emerge as the torchbearer for the burgeoning PC culture much in the same way the *Whole Earth Catalog* had for the DIY back-to-the-landers of the late 1960s and early 1970s.

Once they found Levy's book, Brand and Kelly decided to hold a conference to bring together the three generations of hackers. As Kelly later told Stanford communication professor Fred Turner, he and Brand wanted to see whether hacking was "a precursor to a larger culture" and they hoped to "witness or have the group articulate what the hacker ethic was."

In November 1984, around 150 hackers trekked to Fort Cronkhite, nestled in the Marin Headlands north of the Golden Gate Bridge. In attendance for the weekend-long conference were Apple's Steve Wozniak, Ted Nelson (one of the inventors of hypertext), Richard Stallman (the MIT computer scientist who later founded the Free Software Foundation), and John Draper, aka "Captain Crunch" because he discovered one could make free phone calls by using a toy whistle that came bundled (for free!) in a cereal box. Along with meals and beds, Brand and Kelly provided the hackers with computers and audiovisual equipment.

Two topics continually cropped up in the conversations: how to characterize the "hacker ethic" and what types of businesses were emerging within the computer industry. It was then that Brand restated rule three in a way that would come to define the budding digital age. He said:

> On the one hand information wants to be expensive, because it's so valuable. The right information in the right place just changes your life. On the other hand, information wants to be free, because the cost of getting it out is getting lower and lower all the time. So you have these two fighting against each other.

This is probably the most important—and misunderstood—sentence of the Internet economy.

What's especially important about this quote is that it establishes the economic link between technology and ideas. Moore's Law is about the physical machinery of computing. But information is the weightless commodity on which that machinery acts. Physics determined that a transistor would someday be practically free. But the value of the bits the transistor processed—information—well, that could have gone either way.

Perhaps information would become cheaper, because the bits could be reproduced so easily. Or perhaps it would become more expensive, because the perfect processing of computers could make information of higher quality. In fact, it was exactly this question that led to Brand's comment, which addresses both extremes.

Usually the only part of that quote that is remembered is "information wants to be free," which is significantly different from the original Samson quote on Levy's list in two ways. First, Samson meant "free" as in "unrestricted"—those were the days of the mainframe, and the big issue was who could get access to the machine. Brand, however, had evolved the meaning to the one of this book—free as in zero price.

The second difference is that Brand turned Samson's "should" into "wants to be." Much of the force of Brand's formulation is due to the anthropomorphic metaphor that imputes desire to information, rather than projecting a political stance ("should") upon it. This value-neutral phrasing wrestled "free" away from the hacker zealots such as Stallman, who wanted to protect an ideology of forced openness, and expressed it as a simple force of nature. Information wants to be free in the same way that life wants to spread and water wants to run downhill.

This quote is misunderstood because it is only half-remembered. Brand's other half—"information wants to be expensive, because it's so valuable"—is ignored, perhaps because it seems both paradoxical and tautological. Perhaps a better way to understand it is this:

Commodity information (everybody gets the same version) wants to be free. Customized information (you get something unique and meaningful to you) wants to be expensive.

But even that's not quite right. After all, what is a Google search if not a unique and customized sort of the Web, tailored just for you to be a meaningful response to your query? So let's try again:

Abundant information wants to be free. Scarce information wants to be expensive.

In this case we're using the marginal cost construction of "abundant" and "scarce": Information that can be replicated and distributed at low marginal cost wants to be free; information with high marginal costs wants to be expensive. So you can read a copy of this book online (abundant, commodity information) for free, but if you want me to fly to your city and prepare a custom talk on Free as it applies to your business, I'll be happy to, but you're going to have to pay me for my (scarce) time. I've got a lot of kids and college isn't getting any cheaper.

BRAND EXPLAINS

But that's just my interpretation. Given the impact of his prophecy, I went to Brand directly to better understand the context and meaning as he intended it.

My first questions pertained to the particular phrasing of his legendary remark. First, why did he change the hacker imperative that information "should" be free to "wants to"?

Two reasons, he said. First, from a semantic perspective, it just sounded better: "It's poetical and mythical and it gets away from the finger-wagging 'should.'" But the second reason is more important: "It flips the perspective from yourself to the phenomenon, and the phenomenon is that value is coming from this peculiar form of sharing." In other words, it's more a function of information than it is a decision that you or I make about it. It really doesn't matter what our particular philosophy is about charging for or giving away information, the underlying economics of information clearly favor the second option.

My next question in the deconstruction of this sentence was about the oft-forgotten second part. Why did he construct this duality of "free" and "expensive"?

He said he was drawn to the paradox of information being pulled in both extremes:

> In arguments I was hearing about intellectual property, both sides made perfect sense, and that is the definition of a paradox.

Paradoxes drive the things we care about. Marriage is a paradox: I can't live with her, and I can't live without her. Both statements are true. And the dynamic between those two statements is what keeps marriage interesting, among other things.

Paradoxes are the opposite of contradictions. Contradictions shut themselves down, but paradoxes keep themselves going, because every time you acknowledge the truth of one side you're going to get caught from behind by the truth on the other side.

At the conference there were some people who were distributing free shareware, and others who were selling copy-controlled enterprise software for thousands of dollars. So the price that you could charge for this stuff was still in the process of being discovered, and the price kept going both higher and lower. In other words, the market never cleared in any normal sense. People were charging whatever the traffic would bear, and the traffic put up with all kinds of very weird prices. You could hold corporations up like a total bandit.

Another subtlety in the sentence is his use of the word "information." This is a relatively modern use of the term, which dates to Claude Shannon's famous 1948 paper on information theory. Before that people generally used different words (or no word at all) to describe the particular phenomenon of ideas or instructions encased in code. (Indeed, in his 1939 writings on his emerging ideas, Shannon himself used the word "intelligence" instead.) One of those words people used was, of course, "language," but others included "symbols" and "signs." Until the information age, the word "information" was usually used in the context of news: "I've got some new information." Or simply, "facts."

Shannon worked at AT&T, and his theory was based in a context of signal processing. It defined information as the opposite of noise—coherent versus incoherent signals—and he calculated how to extract one from the other. That can be done in the form of analog or digital signals, but today when we talk about information, we're usually talking about digital bits: those on/off signals that mean nothing or everything, depending on how we decode them.

A word processor thinks your MP3 file is just noise, and your TiVo can't read a spreadsheet, but from an information perspective, they're all the same thing: a stream of bits. A bit reflects just the difference between two states, which may or may not have meaning. But information is what British anthropologist Gregory Bateson described as "a difference that makes a difference."

When Brand used the word "information" he meant digitally encoded information, and what this reflected was his experience with early digital networks, including the one he cofounded, the Whole Earth 'Lectric Link (WELL). What he had learned from them was that the bits and their meaning were entirely different things. The bits were, economically at least, virtually free, but their meaning could have a wide range of value, from nothing to priceless, depending on who was receiving them.

"One of the things that I used as a model of the WELL was the telephone company," he explained. "It does not sell you conversation. They really do not care what anybody says to each other. All they want is to have you pay your bill for having the phone working, and a certain amount of time on it. Content is irrelevant."

The physical world analogy, he said, is a pub. It provides a place for community and conversation, but it doesn't charge for that. It just charges for the beer that lubricates it. "You find that something else to charge for, whether it's the steins of beer or the dial tone, or some other equivalent, like adjacent advertising. You always wind up charging for something different than the information."

And does it annoy him that for twenty-five years, people have been quoting only half of his phrase? That's what happens to memes, he says: They propagate in their most efficient form, whether that was what was intended or not. After all, he notes, Winston Churchill did not say, "Blood, sweat, and tears." Winston Churchill said, "Blood, sweat, toil, and tears." That may sound better, but one of them is not a juice. Mimetic propagation edited the phrase to its optimal form.

7

COMPETING WITH FREE

Microsoft Learned How to Do It Over
Decades, but Yahoo Had Just Months

ON FEBRUARY 3, 1975, Bill Gates, then "General Partner, Micro-Soft" wrote an "Open Letter to Hobbyists," explaining that his new company had spent $40,000 developing software that was being copied for free. If this continued, he warned, he would be unable to develop new software in the future and everyone would lose:

> As the majority of hobbyists must be aware, most of you steal your software. Hardware must be paid for, but software is something to share. Who cares if the people who worked on it get paid?

Eventually, it worked. As the personal computer moved from the geeky world of hobbyists to regular users who were less adept at copying software, the notion that code should be paid for became accepted. Along with the Apple II and IBM PC came the rise of stores that sold software in boxes, complete with instruction manuals. Software became an industry and Microsoft, now without a hyphen, grew rich.

But its days of competing with Free were not over. Piracy never completely went away, and once software moved from hard-to-copy floppy disks to CDs, which could be duplicated the same way music CDs were, it boomed. Microsoft added security codes that users would get in the official packaging, but the pirates just copied them, too, along with the holograms on the packaging. Lawsuits, awareness campaigns, industry trade groups, and even diplomatic action kept piracy in check in the developed world, but in the developing world it ran wild.

In China, the fast-growing PC market was turbocharged by the pirates who sold not just Microsoft's software but everyone else's, too, from games to educational programs. Officially, Microsoft took a hard line against this. But Gates and Co. were realists, too. They knew that piracy of their products was impossible to wipe out entirely, and that any attempt to do so would be both expensive and painful for their paying customers, who would have to jump through all sorts of verification hoops. And it wasn't all bad: If people were pirating the software, they were at least using it, and this mind-share could someday translate into real-market share once these countries developed.

"Although 3 million computers get sold each year in China, people don't pay for our software," Gates said in 1998 to a group of students at the University of Washington. "Someday they will, though, and as long as they're going to steal it, we want them to steal ours. They'll get sort of addicted, and then we'll somehow figure out how to collect sometime in the next decade."

Now that time is coming. China got richer, computers got cheaper (the hottest category is now "netbooks," stripped-down laptops that cost as little as $250), and Microsoft lowered its prices for operating systems on such machines to around $20 (less than a quarter what it charges for the usual versions). Piracy created dependency and helped lower the cost of adoption when it mattered. Today, after a few decades of piracy, you've got a huge paid market in China alongside the pirate one, continued Microsoft dominance, and consumers who have more money and less tolerance for the hassles that come with unauthorized software. Gates's strategy of doing just enough to keep piracy to a dull roar, rather than imposing the brutal things that would have been required to actually eliminate it, paid off.

FREE TRIALS

In the 1990s, while Microsoft was fighting piracy abroad, it was competing with a different kind of Free at home. Having won the operating system wars, it was battling to maintain its lead with applications software, from word processors to spreadsheets. Competitors such as WordPerfect Office and Lotus SmartSuite charged PC makers rock-bottom prices to have their software "bundled" with new computers. The hope was that new PC consumers would use the software that came with the machine, investing in the programs with their learning and files, and when it came time to upgrade to a paid version they'd be hooked.

This slowed Microsoft's market share growth enough to worry Gates. He decided to respond in kind. Microsoft developed its own stripped-down version of Office, called Microsoft Works, which it charged PC makers just $10 to bundle with new computers. This effectively matched the low price offered by competitors, and because Works was file-compatible with full-blown Office, it was a way to keep consumers in the Microsoft sphere of influence, even if the company wasn't making much money from the entry-level product.

This same strategy served Microsoft as the world moved from the desktop to the Web. Netscape released its Web browser, Navigator, for free, instantaneously de-monetizing that nascent industry. What's worse, Netscape's free browser was meant to work best with its own proprietary Web server software, in an effort to cut away at Microsoft's lucrative server operating system market.

Once again, Microsoft was forced to respond. It quickly developed its own free Web browser, Internet Explorer, and bundled it with every version of its operating system. This had the desired effect of checking Netscape's growth, but Microsoft paid the price with a decade of anti-trust prosecutions and fines for anticompetitive behavior. Trust busters

HOW CAN HEALTHCARE SOFTWARE BE FREE?

Since November 2007, thousands of physicians have signed up to receive free electronic health record and practice management software from San Francisco-based start-up Practice Fusion. Enterprise software for medical practices can cost $50,000. How can one company give away its e-record system at no charge?

Sellling data can be more profitable than selling software.

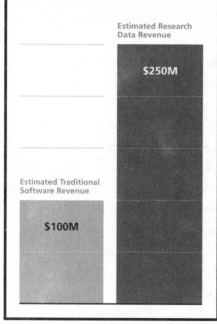

Estimated Research Data Revenue

$250M

Estimated Traditional Software Revenue

$100M

▶ **Freemium + advertising.** Tapping the freemium model, Practice Fusion offers two versions of its software: a free one that serves ads (à la Google AdSense), and an ad-free one that costs $100 per month. Of the first 2,000 doctors to adopt Practice Fusion's e-record system, less than 10 percent opted to pay. But the real revenue lies elsewhere...

▶ **Sell access to your data.** Using free software, Practice Fusion attracts a critical mass of users (doctors) who, in turn, create a growing database of patients. Medical associations conducting research on specific conditions require longitudinal health records for a large set of patients. Depending on the focus of a study (think: white, middle-aged, obese males suffering from asthma), each patient's anonymized chart could fetch anywhere from $50 to $500. A physician typically sees about 250 patients, so Practice Fusion's first 2,000 clients translates to 500,000 records. Each chart can be sold multiple times for any number of studies being conducted by various institutions. If each chart generates $500 over time, that revenue would be greater than if Practice Fusion sold the same 2,000 practices software for a one-time fee of $50,000.

attacked it for "tying" a free product to a paid one. Free is fine, the regulators said, but not if you're a monopoly and are using Free to keep competitors out.

In antitrust theory, because the dominant company in the market has unmatchable ability to subsidize a free product with a paid one (on which they may be extracting monopoly rents), they have to be more careful in how they use Free. In the end, Microsoft was allowed to continue to bundle free software, from the browser to backup utilities, with its operating systems, but the cost was hundreds of millions of dollars in fines, legal fees, and damage to the company's reputation.

THE PENGUIN ATTACKS

Another form of free software is open source, something Microsoft has been competing with for decades, although not always by that name. Until 1998, software that people could use and modify without charge was called "free software" or "freeware," and ranged from operating systems (such as variants of UNIX) to word processors and programming languages. But with the rise of the Web as a communications platform, the informal communities of programmers who write this code became larger and more effective. Special licenses were created that allowed software to spread and attract more contributors. Free software became a force to be reckoned with.

Netscape's 1998 decision to publicly release the Netscape browser code was the catalyst that took free software mainstream. In a meeting convened later that year by publisher Tim O'Reilly, a consensus emerged around the term "open source." This had the main advantage of not using the word "free," which had been muddied by the ideological extremism of Richard Stallman, the former MIT firebrand whose Free Software Foundation had been trying to push the movement toward his own anticapitalist views.

Present at the meeting was Linus Torvalds, then twenty-nine. Seven years earlier, in Helsinki, he had started work on a modest project to create a simplified variation of the UNIX operating system that he called Linux. Due to a combination of good code, his winning personality and organization skills, and, most importantly, the Web as a vehicle of global collaboration, it took off (fear of Microsoft's domination and the general anti-Redmond sentiment of software purists didn't hurt, either).

By the time of the O'Reilly meeting Linux was already seen as the poster child for this new class of software, an example of functional, popular code that was built on a license that required anybody who used and changed the software to make those changes free and open to all. Anybody could sell the software if they wanted to, but they couldn't own it.

Initially, Linux was mostly competing with other variants of UNIX, from the free ones to Sun's and IBM's commercial versions. But its success, both in its technical abilities and its extraordinary harnessing of free talent and labor, was starting to register in Redmond, too, where Microsoft was sitting happily on a multibillion-dollar market for server operating system software.

In interviewing Microsoft executives about how the company figured out how to compete effectively with open source, one of the most startling things for me was how late the dates begin. Although the company had been aware of Linux from the beginning and its marketers had been dismissing it publicly since the late 1990s, within Microsoft it was seen as just another gnat on its hide; not serious enough to cause a shift of strategy. The company pegs Linux World 2002, a conference that was held in September of that year, as the beginning of what program manager Peter Houston calls the "engage Linux credibly" strategy.

As a point of reference, that epiphany was more than a decade after Torvalds started the Linux project and four years after O'Reilly's open source summit. It was three years after the "Linux bubble" of companies such as VA Linux and Redhat that went public on NASDAQ and saw their shares skyrocket. And, by 2002, the Linux market share of the Web server operating system market stood at about 25 percent, compared to Microsoft's 50 percent.

The story of why this took so long and what happened next can best be told through psychiatrist Elisabeth Kübler-Ross's Five Stages of Grief.

STAGE 1: DENIAL

Where had Microsoft been for Linux's first decade? Mostly hoping the free operating system would go away or remain insignificant, like most other free software had to date. Even if it didn't disappear completely, Microsoft executives hoped the appeal of Linux would be mostly to people who already used UNIX, rather than Microsoft's own operating systems. That wasn't entirely reassuring—those UNIX customers were a market Microsoft wanted, too—but it was better than direct competition. But more than anything else, Microsoft managers were confused by why any customer would want free software and all the headaches that came with products not polished to a professional sheen.

But the customers did, especially as they built larger and larger data centers to run the fast-growing Web. Maintaining one Linux server might be harder than its equivalent Microsoft counterpart, but if you're going to deploy hundreds or thousands, learning the quirks of Linux once could save a huge amount of money down the road. By 2003, Linux's share of the Web server market had grown closer to one-third. One way to stem the tide would have been to match the Linux price: zero. But that was simply too scary to contemplate. Instead, Microsoft mostly sniped from the sidelines.

Within the company, some engineers were already warning that Linux represented a long-term competitive threat to Microsoft's core business model and arguing that the company had to mount a more credible response. In 1998, one programmer circulated a memo describing open source software as a "direct revenue and platform threat to Microsoft." The document, which was leaked and circulated as the "Halloween memo" (both for when it was leaked and the scary nature of its contents), goes on to warn that the "free idea exchange in OSS has benefits that are not replicable with our current licensing model and therefore present a long term developer mindshare threat."

But in public, Microsoft was taking a very different stance. One news report from December 1998 goes like this: "Microsoft executives dismiss open-source as hype: 'Complex future projects [will] require big teams and big capital,' said Ed Muth, a Microsoft group marketing manager. 'These are things that Robin Hood and his merry band in Sherwood Forest aren't well attuned to do.' "

STAGE 2: ANGER

Once it became clear that Linux was not only here to stay, but really competing with Microsoft's product, the company turned hostile. Sure, Linux was free, salespeople told wavering customers: "Free like a puppy." Visions of a lifetime of dog food, poop, and twice-a-day walks froze them in their tracks.

Microsoft decided to make economics the attack strategy. The key phrase would be "total cost of ownership." The real cost of software was not its price, but its upkeep. Linux, they argued, was harder to support, and the suckers who went for it would pay every day for the armies of programmers and IT people they'd need to keep this bag of bolts working.

In October 1999, Microsoft took the gloves off and published a document titled "Five Linux Myths." It cataloged technical deficiencies and concluded that Linux's performance didn't stand up to Microsoft products. And free wasn't really free. "Linux system administrators must spend huge amounts of time understanding the latest Linux bugs and determining what to do about them," it warned. "The Linux community will talk about the free or low-cost nature of Linux. It's important to understand that licensing cost is only a small part of the overall decision-making process for customers."

However, it wasn't working—in the absence of proof, customers dismissed this as more Microsoft FUD (fear, uncertainty, and doubt). Linux and other open source software projects such as the Apache Web server, MySQL database, and Perl and Python programming languages continued to gain ground. In November 2002, a frustrated Windows program manager fired off a memo to Microsoft's public relations department: "We need to more effectively respond to press reports regarding Governments and other major institutions considering [open source] alternatives to our products. . . . We must be prepared to respond . . . quickly and with facts to counter the perception that large institutions are deploying [open source software] or Linux, when they are only considering or just piloting the technology."

STAGE 3: BARGAINING

By the time the 2002 Linux World arrived, it was clear within Microsoft that they needed a new strategy. IBM had already created a Linux division and assigned its engineers to start writing code for the project. It was time for Microsoft to turn down its customary stream of vitriol and face the facts: Linux wasn't going away, and customer anger with Microsoft's tactics was part of the reason why. "We realized that we had to take the emotion out of it if we expected anyone to take us seriously," says Houston, who ran Microsoft's team working on competing with Linux. "As it was, everything we said just dug our hole deeper, to the delight of our competition." At Linux World, the Microsoft representatives wore T-shirts that said "Let's Talk."

After the conference, Houston understood why Microsoft hadn't been getting traction. "We needed to *prove* what we'd been saying: that Linux had higher cost of ownership." So he commissioned an independent study by IDC, a consultancy, to find out if Windows really was better than Linux when total cost of ownership was factored in. The results came back as a clear win for Microsoft, but the executives were torn as to whether they should use the report or not. After they had claimed the same with less evidence for so long, would this change anybody's mind?

Perhaps not, but it did win Microsoft a place at the table. Customers realized that Microsoft was not just spinning—Linux really was more complicated and costly than it looked. Meanwhile, Microsoft decided to dip its own toes in the openness waters. It announced a "shared source" program by which government customers could see the underlying code for Windows and other Microsoft products. If one of the appeals of open source was transparency, Microsoft would provide it—but only after swearing the customers to secrecy and otherwise ensuring that the code didn't leak. A few government buyers went through the process, but it hardly made a dent in the Linux moment. It was time to do something more radical.

STAGE 4: DEPRESSION

In late 2003, Microsoft surprised everyone by hiring Bill Hilf, who had run IBM's successful Linux strategy. During the recruiting process Steve Ballmer, Microsoft's CEO, told him, "We have to have an answer to Free." Nothing the company had done so far had stemmed the tide, and when Hilf arrived and started talking to engineers, he could see why. "In my interviews it was clear that they had no idea how open source worked," he said. "There was massive misunderstanding—they saw it as *only* a threat."

One of the reasons Microsoft seemed so ill-informed about open source was that its lawyers had forbidden its engineers from working with it. The license that Linux and similar open source software uses, known as the GPL (general public license), requires that every "derivative work" of open source software also be open source. The lawyers decided that this made it a virus: Any Microsoft programmer who touched it might be at risk of infecting anything else he or she worked on, with a possibility that one mistake could even accidentally open-source Windows.

So when Hilf wanted to build an open source lab at Microsoft, it was treated like a biohazard facility. The buildings department punched a hole in a former storeroom and let him thread network cables through. But after that he was on his own, without a budget. Hilf had to use recycled computers and circulate a "Help Bill" campaign to get spare equipment. Anybody who worked on open source couldn't work on any other Microsoft project, for fear of spreading the GPL disease. The *Seattle Post-Intelligencer* called him "the loneliest man in Redmond."

STAGE 5: ACCEPTANCE

Today Hilf's open source lab is full of humming, high-end servers, purchased new. He has a budget and a staff of programmers working on open source projects. What changed? Pragmatism at the top. Gates and Ballmer had taken their best shot at Linux and it was only getting stronger. It was time for Microsoft to adapt to the new reality. Microsoft's position is now that it has to "interoperate with free," which is to ensure that its software works with open source and vice versa. Its programmers get around the lawyers' fears by only submitting "patches," rather than working on core open source code.

The market share numbers tell the story. Microsoft still has the largest market share in servers, and Linux is still just around 20 percent. In other markets, such as desktop operating systems and office suites, Microsoft's share is closer to 80 percent. The market has decided that there's a place for all three models: totally free, free software and paid support, and good old pay for everything.

Smaller users, from Web start-ups to price-sensitive individuals, often choose open source software, which gets better and better every day. But big companies care more about minimizing risk: They're willing to pay for their software, either from Microsoft or from commercial Linux distributions such as Red Hat, because when they write a check they get a contract. And with that contract comes "service level agreements," which is another way of saying that when things don't work, they've got someone to call.

Today, both open and closed source are huge markets. In dollar terms, Microsoft's revenues dwarf any of its open source competitors. But in terms of users, it's a lot closer. The Firefox browser, for instance, continues to gain on Microsoft's IE (it now has about 30 percent of the market), and the nonprofit company that makes it, Mozilla, funds the browser's development almost entirely with a cut of Google's ad revenues when people use the Firefox search bar, which sends them to Google's search result page. Mozilla's staff is fewer than a hundred people, yet it's running circles around Microsoft's browser team. It's another business built on Free, no tie-in to a commercial operating system required.

Meanwhile, most of the big Web sites, from Google to Amazon, are running primarily on open source software. Even in the most staid companies, open source is creeping in with languages such as Java and PHP.

It's a hybrid world, with free and paid coexisting. The lesson from Microsoft's history is that's not only possible, it's likely. One size doesn't fit all.

CASE TWO: YAHOO VS. GOOGLE

On April 1, 2004, Google issued a press release announcing a new Web mail service, called Gmail. Given Google's track record of gag announcements on April Fool's Day, there was some question as to whether this was for real.

But six miles south of the Googleplex, at the Yahoo headquarters, there was no doubt that Google was dead serious. Yahoo executives had been expecting this day for years, since they first got wind that Google was planning to launch an email product and had registered gmail.com.

Yahoo was by far the largest Web mail provider, with around 125 million users. It was a good business. Most people used the free version, which offered ten megabytes of storage. If people wanted more, they could pay for various premium services from twenty-five megabytes to one hundred megabytes and avoid advertisements. The business was profitable, and Yahoo was increasing its lead over competitors such as Microsoft and AOL.

But in early 2004 the rumors of Google's intention to enter the market were disquieting. It was not just that everything Google touched seemed to turn to gold, but the word was that Google was going to launch with one gigabyte of storage (one thousand megabytes) free—one hundred times what Yahoo offered.

Yahoo executives Dan Rosensweig, Brad Garlinghouse, and Dave Nakayama huddled to consider their options. They had to do something—Google had terrifying momentum and was big enough to take a huge chunk of the email business if it wanted to. And if Gmail was really going to offer a gigabyte for free, it could be potentially ruinous for Yahoo to match.

HOW CAN TRADING STOCKS BE FREE?

If E*TRADE was the first venture to disrupt the stockbroker industry by tapping on-line efficiencies, then Zecco.com represents the next wave. On Zecco.com, investors make up to 10 stock trades per month at no charge. Since the e-broker began offering free trades in 2006, more than 150,000 members have joined. Even as the market plunged in the fall of 2008, account sign-ups increased by 50 percent and the number of daily trades increased by a third. How can Zecco afford to take zero commission from a client a discount broker might charge $100?

How Zecco Makes $179/year from a Semi-Active Trader

$65 - Interest

$50 Tax/Account Management Software

$45 Biannual Portfolio Rebalance

$19 - Options Trades

▶ **Set minimums, charge for additional trades.**
Traders are granted 10 free trades only if they maintain an account balance of $2,500 of total equity. Drop below $2,500 and Zecco charges $4.50. Likewise, every trade after the first 10 costs $4.50. One-fourth of all Zecco customers make more than 10 trades per month (at least $170,000 per month for Zecco). Though typical users make just one to two monthly trades and maintain a balance above $2,500, they may rebalance their portfolios once or twice a year. Each time, they average 15 trades ($45 per year for 45 trades, 10 to 20 of which are free).

▶ **Make money on un-invested funds.**
This is a no-brainer. All online brokers take advantage of this. An average Zecco user might maintain $1,500 of un-invested cash in his account, just in case a promising investment opportunity arises. Like a bank, Zecco makes 2% in interest (in this case, $30/year). If the investor holds a margin balance of $500, Zecco takes 7% (another $35/year).

▶ **Supplement with paid services and ads.**
For clients looking to reduce capital gains taxes, Zecco sells tax planning and portfolio management software. After a two-month trial (which is free), customers pay $25 every six months ($50/year). Active traders also subscribe to market data feeds for $20 per month ($240/year). And like most commercial sites, Zecco runs topical banner ads.

The problem is a classic one in Free. It's easier for the newcomers than for incumbents. That's not just because the incumbents have a revenue stream that they're in danger of cannibalizing. It's also that they have a lot more users, and the costs of serving millions of customers can be astronomical.

Google had no email customers, so it could offer a gigabyte of storage without bearing any real cost: A few servers should handle the first few thousand customers (and, as it turned out, Google would keep the service invitation-only for its first year, ensuring that it could handle the demand without having to buy a lot of hardware). Yahoo, on the other hand, had millions of customers. If it offered the same thing, it might have to buy a warehouse of servers to satisfy the increased demand for email storage.

The more the Yahoo executives thought about it, the worse it looked. Would their premium subscription business, which was bringing in direct revenues, not just advertising, vaporize when people could get ten times as much storage for free? Would people abuse the system, using all the capacity Yahoo offered them as a form of free backup? And, worst of all, they realized that they probably couldn't just match Google—to maintain their lead they would have to offer even *more*.

The executives imagined the building full of "spinning disks"—the most expensive kind of storage, from the hardware to the electricity costs—that they would have to buy just to counter Google's press release. It was depressing—and unfair. But what was their choice?

Garlinghouse and Nakayama sat down to run the numbers. The charts filled whiteboards. There was the cost of storage, which was at least falling. Then there was the expected demand for that storage, which showed a classic Long Tail shape: A few users would consume a lot, while most would consume just a little. But how quickly would that change, and how would people's habits of deleting email after they'd read it last when there was no reason to delete anything?

There were also decisions to be made over the different kinds of storage that Yahoo could spread the email over: fast, slow, and slower yet. Perhaps Yahoo could store older email on cheap, slow storage,

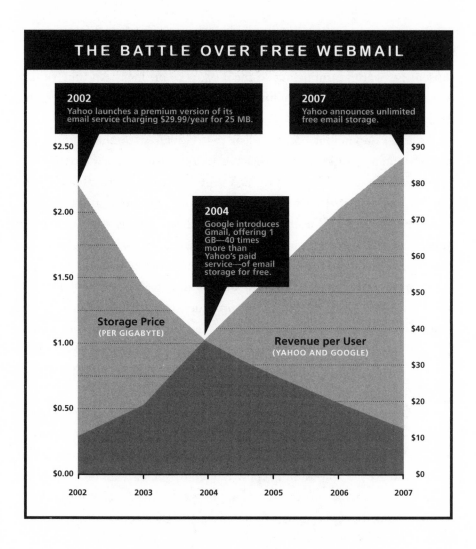

THE BATTLE OVER FREE WEBMAIL

2002
Yahoo launches a premium version of its email service charging $29.99/year for 25 MB.

2007
Yahoo announces unlimited free email storage.

2004
Google introduces Gmail, offering 1 GB—40 times more than Yahoo's paid service—of email storage for free.

Storage Price
(PER GIGABYTE)

Revenue per User
(YAHOO AND GOOGLE)

keeping only the newer email on the more expensive fast storage, where it could be quickly searched for and retrieved. But that would require a whole new email software architecture, which introduced even more cost and risk.

Finally, there was the revenue side. Yahoo's email wasn't just earning the company money by displaying ads and selling premium subscriptions, it was also increasing consumer attachment to the rest of Yahoo. As people went from email to the Yahoo home page or any of its other services, from finance to news, the company made lots more money. Yahoo couldn't afford to lose market share in email, because those users were so important to the rest of the business. And the value of each user was going up along with advertising rates (see chart on page 115).

As 2004 dawned, it became clear that Google was indeed going to release Gmail. Yahoo needed to have its response ready. On April 1, Gmail launched, and it was just as Yahoo had feared: one gigabyte of storage for free. So Rosensweig, then Yahoo's chief operating officer, pulled the trigger and authorized the purchase of tens of millions of dollars' worth of server and storage equipment. On May 15, at an analysts meeting, Yahoo announced that free email storage would go from ten to one hundred megabytes immediately and would soon go higher—premium users who had paid for that much storage could have a refund. By the end of the year it had matched Gmail's one gigabyte, and in 2007 Yahoo went all the way, announcing unlimited email storage for free. (Meanwhile Gmail has only gradually increased its free storage, which now stands at just under eight gigabytes.)

What happened after this surprised all the Yahoo executives. Users didn't flee Yahoo's premium email package in droves. There were still some features worth paying for, such as Web mail without ads, and even those people who didn't renew tended to stick around for a while since they were on an annual plan. People's email behavior didn't change radically, and they continued to delete messages—storage consumption grew more slowly than feared.

Nakayama's team wrote software that caught abuse efficiently and kept the spammers at bay. The definition of "unlimited" storage was also something Yahoo could control. You could add all the email you wanted, but Yahoo would watch if you were adding it too quickly, which is one sign of abuse. As Nakayama puts it, "You can drive as far as you want, but not as fast as you want." That meant that Yahoo could add storage capacity at a slower pace, and the longer it could wait, the cheaper that storage would be.

In the end, it worked: Yahoo didn't lose any significant market

HOW CAN AN EXCLUSIVE CONFERENCE REMAIN PRICEY IF IT'S FREE ONLINE?

One ticket to TED, the invite-only conference on tech, entertainment, and design, costs $6,000. Each year, CEOs, Hollywood elite, and ex-Presidents flock to a resort in California (now Long Beach, after a quarter-century in Monterey) to watch 18-minute presentations given by the likes of Darwinist Richard Dawkins, Sims creator Will Wright, and Al Gore (and occasionally, me). In 2006, after years of exclusivity, TED started broadcasting the talks on its Web site for free. To date, TEDTalks have been viewed more than 50 million times. How can TED give away its crown jewels?

TED Conference, price and attendance

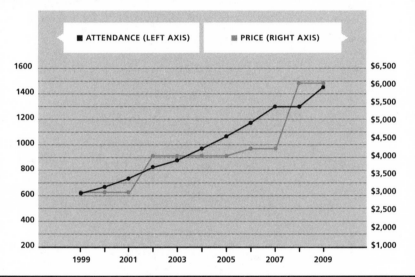

▶ **Streaming content online isn't the same as being there.** Watching the presentations is only part of the experience; an equal part is mingling with other attendees, who are often of the same caliber as those on stage. Come for the talks, stay for the hallway conversations. Plus there's the allure of seeing it first. A ticket to TED isn't devalued by delayed access to the talks; if anything each ticket is worth more now that people know what they're missing. In 2006, the first year TEDTalks were available to anyone with a Web connection, the cost of one ticket was $4,400. By 2008, the price had jumped to $6,000 (double what it was in 1999). Granted the price hike included DVDs and special mailings for members, but let's face it, the ticket is the real incentive. Last year, a ticket was auctioned off on eBay for

charity and sold for $33,850. Sure the auction included a few "perks," like coffee with eBay founder Pierre Omidyar and a lunch date with actress Meg Ryan. But then again, regular TED attendees might do the same; both luminaries are regulars.

▶ **As demand for tickets grows, so does attendance.** Since 1998, attendance at TED has nearly tripled, rising by 10 percent each year. In fact, 2008 was the only year in which attendance did not increase. The reason? The venue in Monterey was simply too small to fit any additional people. In 2009, three years after TEDTalks started broadcasting for free, the conference moved to a theater in southern California with double the capacity.

share. Today, it remains number one and Gmail is a distant number three. Yahoo Mail, rather than turning into a black hole of spending, remained profitable. It competed with Google's Free by becoming even *more* free—getting first to the inevitable end point of unlimited capacity at no cost. Yahoo "rounded down" and it paid off.

But Google wasn't done. Indeed, it had only just begun its march to use Free to enter and compete in any market where software and information economics could disrupt old businesses and create new ones.

8

DE-MONETIZATION

Google and the Birth of a
Twenty-First-Century Economic Model

IT'S NOW BECOME A TOURIST ATTRACTION: 1600 Amphitheatre Parkway, Mountain View, California—the Citadel of Free. This is the Googleplex, the headquarters of the biggest company in history built on giving things away. Outside, surprisingly fit engineers play beach volleyball and ride mountain bikes. Inside, they put their shirts back on and plot new ways to use the extraordinary marginal cost advantages of their huge data centers to break into new industries and expand the search giant's reach.

Today Google offers nearly a hundred products, from photo editing software to word processors and spreadsheets, and almost all of them are free of charge. Really free—no trick. It does it the way any modern digital company should: by handing out a lot of things to make money on a few.

Google makes so much money with advertising on a handful of core products—mostly search results and ads that other sites place on their own pages, sharing the revenues with Google—that it can embrace Free in everything else it does. New services actually start with geek fantasy questions like "Would it be cool?," "Do people want it?," "Does it use our technology well?" They don't start with the prosaic "Will it make money?"

Sound nuts? It might be for GM or GE, but for companies in the pure digital realm, that approach can make perfect sense. Setting out to build a huge audience before you have a business model is not as silly today as it was back in the dot-com era of the late 1990s, when you'd need a wheelbarrow of venture capital cash and racks of Sun servers to do the same. Today any Web start-up has shared access to the same sort of huge server farms that Google uses, which makes the cost of offering services online incredibly cheap. Thanks to the availability of "hosting services," such as Amazon's EC2, that allow companies to launch with no physical infrastructure at all, it's possible to deliver services to millions of users using little more than a credit card.

As a result, companies can start small and aim high without taking huge financial risks or knowing exactly how they will make money. Paul Graham, the founder of Y Combinator, a venture capital firm specializing in small start-ups, gives would-be entrepreneurs simple advice: "Build something people want." He funds companies with as little as $5,000 and encourages them to use open source tools and hosted servers, and to work from their homes. Most use Free to test whether the ideas work and resonate with consumers. If they do, then the next question is what consumers might actually pay for or how else to make money. Years can go by before that day comes (and sometimes it never does), but because the cost of launching the services in the first place is so low, there's rarely huge amounts of capital at risk.

Today there are countless Web companies like this, big and small. But Google is the biggest by far, and because it is so successful in making money in one part of its business, Free is not just an interim step on the way to a business model, it is core to its product philosophy.

To understand how Google became the flagbearer of Free, it helps to see how it evolved. You can shorthand Google's history into three phases:

1. (1999–2001) Invent a way to do search that gets better, not worse, as the Web gets bigger (unlike all previous search engines).

2. (2001–2003) Adopt a self-service way for advertisers to create ads that match keywords or content, and then get them to bid against one another for the most prominent positions for those ads.

3. (2003–Present) Create countless other services and products to extend Google's reach, increasing consumer attachment to the company. Where it makes sense, extend the advertising to those other products, but don't do so at the cost of the consumer experience.

This has worked amazingly well. Today, ten years after its founding, Google is a $20 billion company, making more in profit (more than $4 billion in 2008) than all of America's airlines and car companies combined (okay, that may not be saying much these days!). Not only has it pioneered a business model built around Free, it is inventing an entirely new way to do computing, moving more and more functions that used to run on our desktops into the "cloud," which is to say running in remote data centers and accessed online via our Web browsers (and, ideally, Google's own browser, Chrome).

Where is this cloud? Well, go to another (semisecret) address in The Dalles, Oregon, an area along the Columbia River eighty miles from Portland, and you can see a bit of it, at least from the outside. It's a Google data center—a huge factory-sized building packed with tens of thousands of computer boards and hard drives in racks inside portable containers, all connected together with network wires that ultimately lead to a thick bundle of fiber-optic cables that connect the building to the Internet.

These data centers are the triple play of technology—processing, bandwidth, and storage—embodied. As Google adds more of these information factories around the world, they don't get cheaper, but they do get more powerful. Each new data center's computers are faster than the ones that came before, and its hard drives hold more information. As a result those data centers need bigger pipes to the outside world. Add up all this capacity and you can see why each data factory Google builds can do twice as much for the same price as the one it built about a year and a half earlier.

As a result, every eighteen months the cost to Google of providing you with your Gmail inbox falls by about half. It was only pennies to begin with, but every year it's fewer pennies. Likewise for your directions in Google Maps, your headlines in Google News, and your three-minute entertainment fixes on YouTube. Google keeps building these data centers at the cost of hundreds of millions of dollars, but because the traffic each handles grows even faster than the infrastructure spending, on a per-byte basis the cost to the company of serving your needs falls every day.

Today Google has an estimated half million servers spread out in more than thirty-six data centers, mostly located where electricity is cheap, such as near hydroelectric power stations in the Pacific Northwest.

HOW CAN DIRECTORY ASSISTANCE BE FREE?

AT&T and its competitors rake in $7 billion a year from directory assistance, charging 50 cents to $1.75 per call. Google, on the other hand, offers its automated GOOG-411 service gratis. How can the search juggernaut afford not to charge?

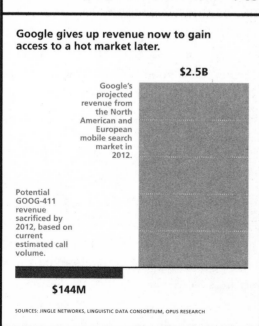

Google gives up revenue now to gain access to a hot market later.

$2.5B

Google's projected revenue from the North American and European mobile search market in 2012.

Potential GOOG-411 revenue sacrificed by 2012, based on current estimated call volume.

$144M

SOURCES: JINGLE NETWORKS, LINGUISTIC DATA CONSORTIUM, OPUS RESEARCH

▶ **Get free data.** Each time GOOG-411 callers request a number, they're giving Google valuable information. Every call provides voice data representing unique variations in accent, phrasing, and business names that Google uses to improve its voice recognition algorithms. Estimated market value of that data since the service launched last spring: $14 million.

▶ **Invest in the next big thing.** The value of voice data hardly compares with potential earnings if Google charged $1 per call. Why give up that cash? Honcho Peter Norvig has said GOOG-411 is a test bed for a voice-driven search engine for mobile phones. If it serves ads to those phones, Google's share of that market could be measured in billions.

(Ironically, electricity is one of the few resources at Google that isn't too cheap to meter; indeed, the electricity consumed by a server now costs more over the life of that server than the server itself. Hence the company's renewable energy program to help invent cheaper, less carbon-intensive energy sources.)

This massive infrastructure is unmatched by any other company in the world, although Microsoft, Yahoo!, IBM, HP, Amazon, and a few others come close. It gives Google great advantages in the Free race. As low as other companies' data costs are, Google's are still lower, and fall faster, because of its economies of scale. By dint of its volume, Google can negotiate the best rates for hardware, bandwidth, and even electricity. (Indeed, CEO Eric Schmidt used to joke that the reason Google's computer racks have wheels on the bottom is so that they can be wheeled out when the owner of the data center goes bankrupt, as happened to every company Google used before it started owning the data centers itself. The pursuit of Free can be a bruising business for those who don't do the calculations right.)

MAX STRATEGIES

Why does Google default to Free? Because it's the best way to reach the biggest possible market and achieve mass adoption. Schmidt calls this Google's "max strategy," and he thinks that such a strategy will come to define information markets. It's very simple: "Take whatever it is you are doing and do it at the max in terms of distribution. The other way of saying this is that since marginal cost of distribution is free, you might as well put things everywhere."

He gives the example of a TV show. Imagine that you and I are the creators of *The Sopranos*. You wonder how we're going to distribute it. As it happens, I've got a friend at HBO and they've agreed to do a weekly series and put up the money to fund it. So that's great, but it's only part of the strategy.

Then we decide we need a blog to build some attention before it airs.

Then, nearer the release, we'll need a PR firm to get some press. Maybe some other online buzz-generators, too, like a Facebook page or some sort of viral video. Then, once the show launches, we'll have a service that will give plot updates via text message and Twitter. This will drive people to the show's Web site, where they can learn more about the characters, which will get them even more interested in the show.

Then we'll take all of the footage we don't put into each Sunday episode and put it on YouTube. Because we generate so much film for *The Sopranos*, we'll have all these extra little scenes. So, to drive even more attention, we'll have a contest for what scene we didn't put in the show that we should have put in the show. And so on. We find a way to take the core idea of *The Sopranos* and distribute it into every possible niche of consumer attention. Maybe only the core one—the HBO deal—makes money, but all the others contribute to its success.

That's a max strategy.

As Schmidt points out, this strategy works great if you've got an HBO. And that is, in a sense, what Google has with its advertising machine (AdWords on search results and AdSense on third-party content). But what if you don't? Then, a max strategy can still earn mass attention and perhaps reputation, but you're left with the challenge of figuring out how to convert that to cash. That's not the worst problem in the world—most companies struggle to achieve popularity, not to monetize it—but if you never quite solve that little detail, "max" can just mean big bandwidth bills for little reward.

Fortunately, this is not Google's problem. It was lucky enough to find a way to make money that grows as fast as Web usage (or even faster, since it keeps gaining market share on its search and advertising competitors). The only thing that limits Google's growth is the pace of growth of the Web itself. So most of its other products are designed, either in part or in whole, to simply extend Internet usage, from free wireless access to free storage.

These other products are what economists call "complements." Complements are products or services that tend to be consumed together, such as beer and salted peanuts, or cars and car loans. For

Google, almost anything that happens online can be seen as a complement to its main business. Every blog post put up is more information for Google's Web crawler to index, to help Google give better search results. Every click in Google Maps is more information about consumer behavior, and every email in Gmail is a clue to our human network of connections, all of which Google can use to help invent new products or just sell ads better.

The interesting thing about the consumption of complementary products is that they tend to rise in tandem. The more people use the Internet, the better it is for Google's core business. So if Google can use Free to encourage people to spend more time online, it will make more money in the end.

Today the vast majority of Google's employees are busy dreaming up new things to give away. It has departments working on ways to give away Wi-Fi and other departments writing open source software. It offers free data storage to scientists and scans classic books to put them online. It gives away photo management software and a place to store those photos online. It freely distributes Google Earth and for good measure has launched satellites to create more imagery to make better maps. It runs a free 411 service that's voice-activated (see the sidebar on page 122). And if you want to create a new cell phone, it will give you the operating system software to make that work, too. No charge.

Schmidt gives an example to explain why such apparent altruism makes sense: "The initial studies on Google News said that people who use Google News were twice as likely to click on search ads on a subsequent search, so everybody said, 'Great.' It's a loss leader—a traffic-getter. Sure, it's a service to the world and so forth, but a more sophisticated view of that is to say that the product is not Google News but Google. It's all about engagement into Google and that if we can get you, at some point in your engagement with Google, to end up using Google for something that we can then monetize, the sums work."

Or, as Nicholas Carr, author of *The Big Switch*, put it, "Google wants information to be free because as the cost of information falls it makes more money."

This is the power of complements.

Because Google's core business is so profitable and is built on such a massive computing infrastructure, it can do everything else it does cheaper and more effectively. It's easier for Google to develop new products, given that they can be built on the work that is already done, and when they're launched it's easier to make them a success thanks to Google's command over global attention. It can introduce products before they're finished ("betas") and quickly get a sense if they're worth pursuing further with a massive trial. Even Google's "failures," such as its Orkut social network or Google Chat, have millions of users. For Google, failure is cheap, so it's not afraid to try risky stuff.

All this sounds very clever, but it's not quite as deliberate as it seems. Although Google does have in-house economists and business strategists, mostly what it has is engineers who are paid to think about what their technology enables and what people might want. Only later does some MBA (a second-class citizen in this geek culture) consider how exactly what the engineers have come up with might be a complement to ad sales.

Sometimes the managers say no on the grounds that the "distraction cost" (the toll it will take on the engineers' other projects) might be too high or that the new creation is not quite as cool as the engineers think, but they never say no just because it won't make money. In the Citadel of Free, gratis is the default. No grand theory is required. It's just the obvious conclusion when you're sitting at the heart of the biggest triple-play cost-reduction machine the world has ever seen.

A GIANT SUCKING SOUND

All this can also sound very scary. While it's great that technology tends to lower prices, it's disruptive when one of those prices is your salary. From the coal miners of Wales to the automotive workers of Detroit, this race to the cheapest, most efficient models has a real human toll. As Jeff Zucker, the head of NBC Universal, put it, the TV industry is terrified of "trading analog dollars for digital pennies." Yet there seems little he or anyone can do to stop it: TV is a scarcity business (there are only so many channels), but the Web is not. You can't charge scarcity prices in an abundant market, nor do you need to, since the costs are lower, too.

It's easy to see why this is scary for the industries that are losing their pricing power. "De-monetization" is traumatic for those affected. But pull back and you can see that the value is not so much lost as re-distributed in ways that aren't always measured in dollars and cents.

To see that at work, look no further than Craigslist, the free classi-fieds site. In the thirteen years since it was founded, its no-charge listings have been blamed for taking at least $30 billion out of America's news-paper companies' stock market valuation. Meanwhile, Craigslist itself generates just enough profit to pay the server costs and the salaries of a few dozen staff. In 2006, the site earned an estimated $40 million from the few things it charges for—job listings in eleven cities and apartment listings in New York City. That's about 12 percent of the $326 million by which classified ad revenue declined that year.

But Free is not quite as simple—or as destructive—as it sounds. Just because products are free doesn't mean that someone, somewhere, isn't making lots of money, or that lots of people aren't making a little money each. Craigslist falls into that second category. Most of the value doesn't go to Craig Newmark, but instead is distributed among the site's hundreds of thousands of users.

Compared to someone placing a classified ad in a print newspaper, Craigslist users save money and can have longer listings. For those browsing the ads, Craigslist offers the usual advantages of the Web, from simple search to automated notifications. Because these two ad-vantages attract lots of people (remember the max strategy), posters are more likely to find a buyer for their apartment or an applicant for their job. And because Free increases the pool of participants, that's likely to be a better apartment, a better job (or applicant) than you'd be able to find in a paid ad equivalent.

Free brings more liquidity to any marketplace, and more liquidity means that the market tends to work better. That's the real reason why Craigslist has taken over so much of the classifieds business—Free attracted people, but the marketplace efficiencies that came with Free ultimately kept them.

"Liquidity" is usually thought of as just a financial term, but in truth it applies in any system of connected parties. In technology, it's called "scale." What it boils down to is that *more is different*. If only 1 percent of the hundred people in some school's sixth-grade class volunteer to help make the yearbook, it doesn't get done. But if just 1 percent of the visitors to Wikipedia decide to create an entry, you get the greatest trove of information the world has ever seen. (In fact, it's closer to one in *ten thousand* Wikipedia visitors who are active contributors.) More is different in that it allows small percentages to have a big impact. And that makes more simply better.

The point is that the Internet, by giving everybody free access to a market of hundreds of millions of people globally, is a liquidity machine. Because it reaches so many people, it can work at participation rates that would be a disaster in the traditional world of non-zero marginal costs. YouTube works with just one in a thousand users uploading their own videos. Spammers can make a fortune with response rates of one in a million. (To give you some context, in my business of magazines, a response rate of less than 2 percent on direct-mail subscription offers is considered a failure.)

For all the cost advantages of doing things online, the liquidity advantages are even greater. There are huge pools of underexploited supply out there (good products and services that aren't as popular as they should be) and equally huge pools of unsatisfied demand (wants and needs people either have but can't act on or didn't even know they had). Businesses such as Craigslist serve to connect them. It's because they can do so cheaply at such massive scale (Craigslist users create more than 30 million classified listings each month, tens of thousands of times as much as the largest newspapers) that they're so successful.

And yet Craigslist makes very little money, just a tiny fraction of what it erased from the newspaper coffers. Where does the wealth go?

To follow the money, you have to shift from a basic view of a market as a matching of two parties—buyers and sellers—to a broader sense of an ecosystem with many parties, only some of whom exchange cash directly. Given the size of Craigslist today (50 million users every month), it's easy to see how more money can change hands there than did in any newspaper classifieds section, leading to better supply/demand matching and economic outcomes for the participants, even though less money remains in the marketplace itself. The value in the classifieds market was simply transferred from the few to the many.

Venture capitalists have a term for this use of Free to shrink one industry while potentially opening up others: "creating a zero billion dollar business." Fred Wilson, a partner at Union Square Ventures, explains it like this: "It describes a business that enters a market, like classifieds or news, and by virtue of the amazing efficiency of its operation can rely on a fraction of the revenue that the market leaders need to operate profitably."

Another venture capitalist, Josh Kopelman, tells this story of one such example:

My first company, Infonautics, was an online reference and research company targeting students. While I was there, I got a firsthand education on "asymmetrical competition." In 1991, when we started, the encyclopedia market was approximately a $1.2 billion industry. The market leader was Britannica—with sales of approximately $650 million, they were considered the gold standard of the encyclopedia market. World Book Encyclopedia was firmly ensconced in second place. Both Britannica and World Book sold hundreds of thousands of encyclopedia sets a year for over $1,000.

However, in 1993, the industry was permanently changed. That year Microsoft launched Encarta for $99. Encarta was initially nothing more than the poorly regarded Funk & Wagnall's Encyclopedia repackaged on a CD—but Microsoft recognized that changes in technology and production costs allowed them to shift the competitive landscape. By 1996 Britannica's sales had dropped to $325 million—about half their 1991 levels—and Britannica had laid off its famed door-to-door sales staff. And by 1996 the encyclopedia market had shrunk to less than $600 million. In that year, Encarta's U.S. sales were estimated at $100 million.

So in just three years, leveraging a disruptive technology (CD-ROM), cost infrastructure (licensed content versus in-house editorial teams), distribution model (retail in computer stores versus a field sales force) and pricing model ($99 versus $1000), the encyclopedia market was cut in half. More than half a billion dollars disappeared from the market. Microsoft turned something that Britannica considered an asset (a door-to-door sales force) into a liability. While Microsoft made $100 million it shrunk the market by over $600 million. For every dollar of revenue Microsoft made, it took away six dollars of revenue from their competitors. Every dollar of Microsoft's gain caused an asymmetrical amount of pain in the marketplace. They made money by shrinking the market.

And now Wikipedia, which costs nothing, has shrunk the market again, decimating both the printed and the CD-ROM encyclopedia markets. (In 2009, Microsoft killed Encarta altogether.) Wikipedia makes no money at all, but because an incomparable information resource is now available to all at no cost, our own ability to make money armed with more knowledge is improved.

The value that Britannica created could once be calculated as some combination of Britannica's direct revenues and the increased productivity of those lucky enough to own the volumes. Wikipedia, being free and easy to access, huge, and otherwise more useful for more people, is increasing the productivity of many more workers than Britannica did. But it isn't making a penny directly; instead, it's taking many pennies away from Britannica. In other words, it's shrinking the value we can measure (direct revenues), even as it's hugely increasing the value we can't (our collective knowledge).

This is what Free does: It turns billion-dollar industries into million-dollar industries. But typically the wealth doesn't vaporize, as it appears. Instead, it's redistributed in ways that are hard to measure.

In the case of classifieds, newspaper owners, employees, and shareholders lost a lot while the rest of us gained a little. But there are a lot more of us than there are of them. And it's entirely possible that the lost $30 billion in newspaper market capitalization will eventually show up as far more than that in increased GDP, although we'll never be able to make that connection explicitly.

Companies that embrace this strategy aren't necessarily calculating the totals of winners and losers. Instead, they're just doing what's easiest: giving people what they want for free and dealing with a business model only when they have to. But from the outside, it looks like a revolutionary act. As Sarah Lacy put it in *Business Week,* "Think Robin Hood, taking riches from the elite and distributing them to everyone else, including the customers who get to keep more of their money and the upstarts that can more easily build competing alternatives."

You see this all around you. Cell phones, with their free national long distance, have de-monetized the long-distance business. Do you see anyone (other than long-distance providers) complaining? Expedia de-monetized the travel agent business, and E*TRADE de-monetized the stockbroker business (and paved the way for other free-trading companies, including Zecco—see the sidebar on page 113). In each case the winners far outnumber the losers. Free is disruptive, to be sure, but it tends to leave more efficient markets in its wake. The trick is to ensure you've bet on the winning side.

THE COST OF FREE

But what if it's not quite as equal as that? What if the wealth is not neatly transferred from the few to the many, allowing a thousand flowers to bloom? What if it really just disappears or, even worse, leads to even fewer winners than before?

This is what Google's Schmidt worries about. The Internet is a prime example of a market dominated by what economists call "network effects." In such markets, where it's easy for participants to communicate with one another, we tend to follow the lead of others, resulting in herd behavior. Since small differences in market share can get amplified into big ones, the gap between the number one company in any sector and the number two and beyond tends to be great.

In traditional markets, if there are three competitors, the number one company will get 60 percent share, number two will get 30 percent, and number three will get 5 percent. But in markets dominated by network effects, it can be closer to 95 percent, 5 percent, and 0 percent. Network effects tend to concentrate power—the "rich get richer" effect.

Although this was the argument used to justify the antitrust prosecution of Microsoft in the 1990s, in this case Schmidt's concern is not about lasting monopolies. In today's Web market, where the barriers to entry are low, it's easy for new competitors to arise. (That, of course, is the argument Google uses to defend itself against charges of being a monopolist.) Nor is it about limited choice: Those same low barriers to entry ensure that there are many competitors, and all the smaller companies and other inhabitants of the Long Tail can collectively share a big market, too. Instead, this is simply a concern about making money: Everyone can use a Free business model, but all too typically only the number one company can get really rich with it.

Why should Google care about whether other companies can use Free to economic advantage? Because it needs those other companies to create information that it can then index, organize, and otherwise package to create its own business. If digital Free de-monetizes industries before new business models can re-monetize them, then everyone loses.

Just consider the plight of the newspapers. The success of the free Craigslist has caused the big city dailies to shrink, taking many professional journalists out of circulation. But low cost and user-generated "hyperlocal" alternatives have not risen equally to fill the gap. Maybe someday they will, but they haven't yet. That means that there is less local news for Google to index. There may be more local information, but it can no longer use the fact that it came from a professional news organization as an indicator of quality. Instead, it has to figure out what's reliable and what's not itself, which is a harder problem.

So Google would very much like the newspapers to stay in business, even as the success of its own advertising model in taking market share away from them is making that more difficult. This is the paradox that worries Schmidt. We could be at a moment where the short-term negative consequences of de-monetization are felt before the long-term positive effects. Could Free, rather than making us all richer, instead make just a few of us superrich?

From the billionaire boss at the Citadel of Free, this may seem like an ironic observation, but it's important to Google that there are lots of winners, because those other winners will pay for the creation of the next wave of information that Google will organize.

"Traditionally, markets are segmented by price, making room for the high-end, the middle, and the low-end producers," Schmidt explains. "The problem with Free is that it eliminates all the price discrimination texture in the marketplace. Rather than a range of products at different prices, it tends to be winner-take-all." His worry, in short, is that Free works all too well for him, and not well enough for everyone else.

Of the richest four hundred Americans, a list that *Forbes* creates each year, I count just eleven whose fortunes were based on Free business models. Four of those, including Schmidt, came from Google. Two came from Yahoo! Two more came from Broadcast.com, an early Web video company that sold to Yahoo! at the peak of the dot-com bubble and whose founders, Mark Cuban and Todd Wagner, subsequently invested well. Then there is Mark Zuckerberg of Facebook and, if you will, Oprah Winfrey, whose $2.7 billion was built on free-to-air TV.

I didn't include all the media tycoons, from Rupert Murdoch to Barry Diller, because they run diversified conglomerates that are a mix of Free and Paid. And the *Forbes* list stops before including a lot of people who have become rich, but not megarich, from the Free model, such as the

MySpace founders and a few open source software heroes such as the founders of MySQL (sold to Sun in 2008 for $1 billion). But Schmidt's point holds: If we measure success in terms of the creation of vast sums of wealth spread among more than a few people, Free can't yet compare to Paid.

But there are signs that this is changing. To see how, you have to look at the evolving nature of the original Free business: media.

9

THE NEW MEDIA MODELS

**Free Media Is Nothing New. What *Is*
New Is the Expansion of That Model to
Everything Else Online.**

IT WAS 1925—the dawn of the commercial radio industry. The wireless craze had swept America, gathering families around the electronic hearth and creating "distance fiends," listeners who marveled at their ability to listen in on transmissions from cities hundreds or thousands of miles away. The miraculous ability of broadcast to reach millions of people simultaneously was forcing radio stations to invent what would become mass media—entertainment, news, and information of the broadest possible appeal. It was the beginning of twentieth-century pop culture. There was only one problem: Nobody had any idea how to pay for it.

Up until then, radio programming was either done on a shoestring (some regional stations would let anyone who walked in the door go on the air) or paid for by the radio receiver manufacturers themselves. David Sarnoff, vice president of the Radio Corporation of America (RCA), explained at the time that "we broadcast primarily so that those who purchase RCA radios may have something to feed those receiving instruments with." But as radio spread, it had become clear that the insatiable demand for new content could not be fed by a few manufacturers alone.

Radio Broadcast magazine announced a contest for the best answer to the question "Who is to pay for broadcasting and how?" Eight hundred people entered, with ideas that ranged from volunteer listener contributions (hi, NPR!) to government licensing and, cleverly, charges for program listings. The winning entry sought a tax on vacuum tubes as an "index of broadcast consumption." (That is, in fact, the model that was adopted in the United Kingdom, where listeners and viewers pay a yearly tax on their radios and TVs and get the ad-free BBC in return.)

There were some suggestions that advertising might be the answer, but it was by far from a popular solution. It seemed a shame to despoil this new medium with sponsored messages. One article fretted that "bombastic advertising . . . cuts into the vitals of broadcast . . . by creating an apathetic public, impairing listener interest and curtailing the sales of receiver sets."

But NBC, one of the new broadcasting companies, was determined to test to see if radio advertising worked. In 1926, it appointed Frank Arnold, the best-known proponent of radio advertising, as its director of development. Arnold described radio as "the Fourth Dimension of Advertising," beyond the prosaic three dimensions of newspapers, magazines, and billboards. Others talked about how radio magically allowed advertisers to become a "guest in a listener's home."

One of the problems with radio, however, was that for all its defiance of distance, distance was starting to strike back. The megastations on the East Coast were using more and more powerful transmissions to reach thousands of miles into the country, but as regional and local radio grew up, the closer local signals were drowning out the national ones. (The Federal Communications Commission was created in part to bring order to the airwaves.) As a result, radio seemed relegated to local advertising, which wasn't lucrative enough to feed all the demand for content.

Salvation came in the form of AT&T, the telephone company. William Peck Banning, later an AT&T vice president, recalled that in the early 1920s "nobody knew where radio was really headed. For my own part I expected that since it was a form of telephony, we were sure to be involved in broadcasting somehow." That turned out to be long-distance transmission of radio programs on AT&T's wires, free of interference, so they could be rebroadcast on local towers around the country. And thus was born the national radio networks, and the first national market for broadcast advertising. (Before then it was limited to smaller pools of local advertising for companies within the range of individual stations.)

A few decades later television followed the same path. Both radio and TV were free-to-air and advertising-supported. It was the beginning of the so-called media model of Free: a third party (the advertiser) subsidizes content so that the second party (the listener or viewer) can get it at no charge.

Today this three-party model is at the core of the $300 billion advertising industry. It not only supports free media, such as traditional over-the-air broadcast, but also subsidizes most paid media, from newspapers and magazines to cable TV, allowing them to be much cheaper than they would otherwise be. And now, with the Web, a medium on which media has no privileged position, it supports everything else.

ADS BEYOND MEDIA

What's different about advertising when it moves beyond media to support software, services, and content created by regular people, not just media companies? Lots. For starters, the usual rules of trust are reversed. I'll give you an example from my own world.

A while back, a friend from Google was visiting our offices at *Wired*. I was showing him our "magazine room," which is where we put all the pages of the issue we're working on in rows on the wall. As the pages take shape, we can move them around on the wall to find the best flow and rhythm for the issue and avoid unfortunate clashes between stories or art elements.

One of the other things we do on this wall is watch for "ad/edit con-flict," which is to say advertisements that appear related to the content they're running against. This stems from the "Chinese Wall" that most traditional media build between their editorial and advertising teams, to ensure that advertisers cannot influence the editorial. But that's not enough. We also need to inspire trust in the readers, so we avoid even the appearance of influence by ensuring that a car ad is not next to a car story or a Sony ad anywhere near our reviews of Sony products. Ideally, we don't even have them in the same issue.

As I was explaining this to my friend from Google, he stared at me with mounting disbelief. As well he should have, because Google does just the opposite.

The appeal of Google's phenomenally successful AdSense program is that it matches ads with content. People pay Google lots of money to do exactly what we forbid: put Sony ads next to Sony reviews. And readers love it—it's called relevance.

Why is such matching bad in print but good online? At the heart of that question is the essence of how advertising is changing as it moves online.

My own, somewhat inadequate, explanation is that people bring dif-ferent expectations to the online world. Somehow, intuitively, they un-derstand everything that my Google friend and I pondered as we stood in that room surrounded by paper. Magazines are put together by peo-ple, and people can be corrupted by money. But Web advertising is placed by software algorithms, and somehow that makes it more pure.

This is, of course, a fiction. Lots of Web ads are placed by hand, and it's all too easy to corrupt an algorithm. But when it's Google placing the ad on somebody else's content, the connection between the two is so arm's-length that people seem not to worry about undue influence.

It's also entirely possible that we in the traditional media business have it all wrong. Perhaps we are just flattering ourselves with our church-and-state pursuit of purity, and readers don't care or even no-tice if a Sony ad is next to a Sony review. Perhaps they would even prefer that and it is our writers who are the real obstacles, afraid that anyone might think that their opinion had been bought. I don't know, but I do know that our industry trade association has strict rules about this sort of thing, and if I dare to break them my magazine will become ineligible for awards and suffer other such punishments.

But what's clear is that the nature of the advertisement is different online. The old broadcast model was, in essence, this: Annoy the 90 percent of your audience that's not interested in your product to reach the 10 percent who might be (think denture ads during football games).

The Google model is just the opposite: Use software to show the ad only to the people for whom it's most relevant. Annoy just the 10 percent of the audience who isn't interested to reach the 90 percent who might be.

Of course it doesn't always work that way, and you've no doubt seen plenty of annoying ads served by Google. But as the increased supply of narrowly targeted ads meets the increased demand from narrowly targeted content, the matching is getting better. For example, on my aerial robotics site, where we run Google AdSense delivering hyper-narrow ads for such arcane products as "three-axis accelerometers," I polled our readers to ask them if they wanted me to remove the ads.

The majority asked me to keep them, because they were so relevant that they counted as content. A smaller group hadn't noticed that there were ads at all. The smallest group of all wanted them gone. (I kept them.)

HOW NEW MEDIA CHANGES OLD MEDIA

One of the interesting things about the ad-supported Free model is that it was actually on the decline in the traditional media business. As television moved from free over-the-air broadcast to cable, which is paid, content was increasingly supported by a mélange of revenue streams, including syndication and cable license fees that had little to do with advertising. Even radio, in the form of satellite radio, was moving to a mix of direct subscriptions and advertising. It was starting to look more like the print media business, which mixes subscription and newsstand sales with advertising revenues.

But the rise of the Web reversed that. After a few years of online experiments with asking people to pay for content, it became clear to almost everyone that fighting digital economics wasn't going to work, and Free won. Not only that, but the price expectations set online began to leak offline, too.

Newspapers realized that the Google generation might not adopt their parents' habit of paying for a print daily, so they introduced free newspapers aimed at young adults and handed out on street corners and subways. Other newspapers kept their price but bundled free giveaways, from silverware to music (see the sidebars on pages 141 and 155). As the rest of the newspaper industry declined, free newspapers became a solitary beacon of hope, growing 20 percent a year (mostly in Europe) and accounting for 7 percent of total newspaper circulation in 2007.

Meanwhile, broadcast television viewership seems to have peaked, at least with the sought-after eighteen-to-twenty-four-year-old viewers, who are increasingly watching clips or even full shows for free online, on YouTube or Hulu. Broadband is the new free-to-air broadcast, and the premium cable lock on the viewer now appears to be eroding.

THE END OF PAID CONTENT

This shift is part of a greater devaluation of content, driven not just by generational taste but also by technological trends. Jonathan Handel, an entertainment lawyer (and former computer scientist) in Los Angeles, gives six reasons for the migration to Free, which I'll paraphrase as follows:

1. **Supply and demand.** The supply of content has grown by factors of million, but demand has not: We still only have two eyes, two ears, and twenty-four hours in the day. Of course, all

HOW CAN SILVERWARE BE FREE?

Controlinveste is one of Portugal's leading media companies, with newspaper, TV, radio, magazine, and Web ventures. Two of its papers boast the country's highest circulations: *Global Notícias*, which is free, and *Jornal de Notícias*, which is paid. As with many paid European newspapers, most sales of *Jornal de Notícias* are via newsstand, where publishers have to win over customers every day. So giveaways are frequently used as marketing devices (as with the free Prince CD included with the *Daily Mail;* see sidebar on page 155). Controlinveste, however, has taken this further than most.

In 2008, Controlinveste gave away a free, 60-piece silverware set with *Jornal de Notícias* to celebrate the paper's 120th anniversary. Every Monday through Friday, one utensil was bundled with the paper. Each Saturday, you'd get a serving utensil (12 total). Miss a day and you're missing a fork or spoon for your set. Silverware was provided in individual packaging to the newsstands, which added them to the papers. It was a hit: Circulation increased by 36% in just three months.

In a mature industry where paid circulation is dropping every year, these results are extraordinary. But how is it profitable? Two ways: silverware is a lot cheaper than we think, especially when purchased in bulk, and secondly, the marginal profit on additional papers sold over the standard base is a lot higher, too. At the newsstand *Jornal de Notícias* costs 0.88 euros or about $1.32 (average from Monday to Saturday). Considering taxes, printing and distribution costs, and the newsstand profit, the newspaper is healthily profitable and all the fixed costs (staff, buildings, and other facilities) are covered between circulation revenues and advertising. But if you sell more copies, it spreads the fixed costs over a larger base, increases the audience, and improves profit margins. How to sell more copies? Give something away!

At the volume Controlinveste buys silverware from China, each piece's cost is measured in just a few cents. When the company includes a spoon with every paper, it eats up much of its profit margin, but eventually, after its fixed costs are covered, the economics brighten since the marginal costs are so low. If giveaways grow circulation consistently, they can promise advertisers a larger audience and charge them more, too.

The free gifts didn't end there. In 2008 alone, Controlinveste also gave away the following:

▶ **Free tool box with tools.** The box was delivered with the Sunday paper (the best-selling and highest-priced edition). A new tool was included every Monday to Friday (177 pieces in all). Result: 20% circulation gain in three months.

▶ **Free DVDs, every Saturday.** You'd pick up a coupon with your Friday paper, then get the movie when you bought your Saturday paper. Result: 47% circulation increase over two months.

▶ **Free dinner plate set.** You'd get a coupon on Saturday and a plate on Sunday, meaning you had to buy the week's two most expensive editions. A complete set was 19 plates. Result: 70% circulation increase over four months.

▶ **Free language course.** A multimedia program for learning English, Spanish, Chinese, French, Russian, Italian, German, Arabic, Greek, Japanese, and Hebrew was given out in parts. You'd get one CD-ROM or book daily for a total of 48 disks, 22 books, and two boxes. Result: 63% circulation increase over 4 months.

content is not created equal and Facebook pages can't compare to the *New York Times*—unless that Facebook page is your friend's, in which case it may be far more interesting than the *Times* (for you). The difference is that there are a lot more Facebook pages than there are *Times* pages, and they're created with no expectation of pay.

2. **Loss of physical form.** We can't help it: We value atoms more
   than bits. As content moved from disks in boxes to files flying
   through wires, it became intangible, even abstract. Plus stealing a
   physical thing deprives someone else of it and costs somebody
   real money—not so for a digital file.

3. **Ease of access.** It's often easier to download content than it is
   to find it and buy it in stores. As such "search costs" decline, so
   does our willingness to pay for having content made available.

4. **The shift to ad-supported content.** Habits set on the Web
   carry over into the rest of life. If content is free online, shouldn't
   it become free elsewhere, too?

5. **The computer industry *wants* content to be free.** Apple
   doesn't make its billions selling music files, it makes it selling
   iPods. Free content makes the devices it plays on more valuable,
   as the radio industry knew back in 1920.

6. **Generation Free.** The generation that has grown up with
   broadband has digital economics somehow wired into their
   DNA. Whether they've ever heard of "near-zero marginal cost"
   or not, they intuitively understand it. That's why they're either
   indifferent or hostile to copyright. They just don't see the
   point.

This is why ad-supported models won online, and why they'll continue
to win.

At this point, the skeptical reader should be on full alert. Surely there
are limits to the advertising dollars out there. Advertising can't support
everything.

This is true, and indeed some advertising can be worth even less online than offline. The reason comes back to scarcity and abundance. As
Scott Karp, the founder of Publish2, a news service and analysis firm,
puts it, "Advertising in traditional media, whether newspapers, magazines, or TV, is all about selling a scarce resource—space. The problem
is that on the Web there's a nearly infinite amount of space. So when
traditional media companies try to sell space online the same way they

sell space offline, they find they only have a fraction of the pricing power."

A glossy print magazine can charge an advertiser more than $100 per thousand print readers, but would be lucky to get more than $20 per thousand online readers. There's simply more competition online— advertisers have more choices and the price falls to whatever the market will bear. But that's for "display advertising," banners and images that are meant to promote a brand, not necessarily to lead to an immediate sale.

There is another kind of advertising, epitomized by Google's text ads that it runs next to search results and on third-party sites. Advertisers only pay when readers click on the ad. Google doesn't sell space. It sells users' intentions—what they've declared they're interested in, in the form of a search query. And that's a scarce resource. The number of people typing in "Berkeley dry cleaner" on any given day is finite.

The result is that while traditional advertising is limited online, the way Google has redefined advertising—connecting products with expressed desires—is still growing fast. Eric Schmidt, Google's CEO, has estimated that the potential market for online advertising is $800 billion, or twice the total advertising market, online and off, today. It's easy to see why: Companies only pay for results. If you're sure to make a dollar for every 10 cents you spend on marketing, the sky's the limit. Compare that with the old Madison Avenue truism: "Half my advertising is wasted, but I don't know which half." No contest.

THE TRIUMPH OF THE MEDIA MODEL

This is why the ad-driven model has spread so far beyond media online. It is simply where the money is. Fred Wilson, the New York venture capitalist, thinks that "most Web apps will be monetized with some kind of media model. Don't think banner ads when I say that. Think of all the various ways that an audience that is paying attention to your service can be paid for by companies and people who want some of that attention."

You can look at the Web as the extension of the media business model to an unlimited range of other industries. Google is not a media company by any traditional definition of the word, but it makes its billions from the media business model. So, too, for Facebook, MySpace, and Digg. All of them are software companies at the core. Some organize other people's content, others provide a place for people to create their own content. But they don't create or distribute content the way a traditional media company does.

But when people think of the "media business model," they usually just think of advertising. That's a big part of it, to be sure, but as those of us in the media business know, it goes far beyond that.

First of all, advertising's move online has created scores of new ad forms beyond the traditional "impressions" model of payment per thousand viewers or listeners. (This is known as "cost per thousand" or CPM, confusingly using the roman numeral M for thousand.) Online variants include "cost per click" (CPC), which is what Google uses, and "cost per transaction" (CPT), where advertisers only pay when a viewer becomes a paying customer, such as in Amazon's Associates program.

Then there is "lead generation," where advertisers pay for the names and email addresses of people who have been attracted by free content, or for information about those consumers. Advertisers can sponsor an entire site or department for a fixed sum, not determined by traffic. They can pay to be included in search results, which Google and others offer. Or they can turn to good old product placement and pay to have their brand or goods included in a video or game.

Add text, video, animation, audio, and virtual world (video game) versions of all of these, and you can see how much the advertising world has changed as it's moved online. Twenty years ago advertising could be broken down into five big categories: print (display and classified listings), TV, radio, outdoor (billboards and posters), and handouts (fliers, etc.). Today there are at least fifty different models online, and each one is changing by the day. It's head-spinning—and exhilarating—to watch an industry reinvent itself in the face of a new medium.

THE ORC ECONOMY

We think of "media" as being radio, television, magazines, newspapers, and journalistic Web sites. But media is really just content of any sort, and the best way to measure its impact in our society is by how much time people spend with it. Measured that way, few of the above can compete with a form of content that we rarely think of as media at all, even though it competes with media directly for attention. That form of content is video games, from Xbox 360 shooters to PC online multi-player worlds.

Not only has the games industry risen to rival Hollywood in sales and television in consumer time, but it is transforming at a much quicker pace. No business is racing to Free faster than the video game business.

Once upon a time people bought video games in stores. They came in boxes and typically cost $40 or $50. You'd bring them home, insert the disk, and play them for a week, and then rarely again. Virtually all of a game's sales would be in the first six weeks after its release. It was like Hollywood, but without the lucrative follow-on DVD and syndication market—a hit-driven business with no second chances for the misses.

Actually, this is still the main way people buy video games, crazy as it may sound. But games are one of the last digital products that are still mostly sold that way, and that model is nearing its end. Just as music and computer software is becoming primarily an online market, so will games. And once you switch from shipping atoms (plastic boxes and disks) to transmitting bits, Free becomes inevitable. Over the next decade, this $10 billion industry will shift from primarily a traditional packaged goods business to an online business built on entry prices of zero.

The first signs of this started to emerge in Asia around 2003. Because software piracy in the markets of China and South Korea had made it difficult to sell games the usual way, game-makers turned to the fast-growing online market instead. Cybercafes were booming in China, bringing the Internet to a population that could not, for the most part, afford computers in their homes. In South Korea, "PC baangs," or computer gaming parlors, started to replace the usual arcades as a place for teenagers and young people to hang out in a country where most still live at home with their parents before marriage.

For players, the advantage of online games is in the quality and diversity of the competition: You're playing against real people, not just prescripted artificial intelligences. In one of the most popular categories, known as massively multiplayer online games (think World of Warcraft or its predecessors, such as Everquest), the games are never-ending and can become an obsession that consumes players for years. Not for nothing is the current champion often called "World of War*crack*."

For the game-makers, the advantages of going online are many. Rather than printing disks, manuals, and boxes and then getting a retailer to stock them, they can just let gamers download the software. That saves a huge amount of money in manufacturing and distribution. Going online opens up unlimited "shelf space," so older and niche games aren't driven out by the newer and more mainstream titles, since they're all equally accessible online. And it provides an easy way to update the software to add features and remove bugs.

But the most important reason games are moving online is that it's a better way to make money. It allows the makers to shift from a hit-or-miss "point of sale" revenue model to one based on an ongoing relationship with the player, just as the Gillette disposable razor blade moved the shaving business from the sale of razors to a lifetime sale of blades.

As a result, the online games industry has become the most vibrant experiment in Free in the world. Including both games that are just distributed online and those that are actually played online, this industry was worth an estimated $1 billion in the United States in 2008, and in China it's even bigger—on track to hit $2.67 billion in 2010. It includes everything from iPhone games that you can download from iTunes (free, paid, or a hybrid of the two), "casual games" played online such as poker and Sudoku, children's games such as Club Penguin, Neopets, and Webkinz, and the booming massively multiplayer worlds.

Each one of these markets has become a petri dish of new forms of Free, and as a result this has become the industry to watch for innovative new business models, many of which have applications outside the world of games. Free is nothing new to games, of course: The basic freemium model has long been a staple of the games industry in the form of limited demos, which are distributed free in games magazines or online and allow you to play a few levels without charge. If you like what you see, you can either buy the full version or pay for a code that will unlock the rest of the levels in the version you've got. But the past few years have seen an explosion of more innovative business models built around Free that have only been possible with ubiquitous broadband Internet access. Here are the five most successful categories:

1. SELLING VIRTUAL ITEMS

In 2008, Target sold more than a million dollars' worth of a plastic card that, to most of its customers, was completely baffling. All it had was a numeric code that worked with something called Maple Story, and it sold in increments of $10 and $25. What's Maple Story? Just ask a twelve-year-old kid (or the parent of one). It's an online multiplayer game that became a breakout hit in its native South Korea, where it is played by more than 15 million people, and was imported into the United States by its maker, Nexon, in 2005. Today it has more than 60 million registered users globally.

Like many online multiplayer games, Maple Story is free to play; you can happily move through the levels, interact with other players, and otherwise have fun without spending a penny. But if you'd like to do it *faster,* you may want to buy a "teleportation stone," which will allow you to jump from place to place rather than trudging through the landscape. For that you'll need "mesos" (credits) that you can either earn or get from this card (or, if you're an adult and have a credit card, buy online).

Similarly, Maple Story will let you buy virtual items that will allow you to collect mesos more quickly or move between worlds without having to wait for a "bus." You can buy a "guardian angel" who will bring you back to life immediately, without having to trudge back from a respawn point. With your Nexon points you can buy new outfits, hairstyles, and faces. Importantly, you can't buy a superweapon, because that would be unfair—the company doesn't want people to be able to buy their way to power, creating a two-tiered society. Instead, money is used to save time, look cooler, or otherwise do more with less effort. The opportunities to pay are "nonpunitive," says former Nexon North America boss Alex Garden. You don't have to pay, but you may want to.

But the biggest example of this market in virtual goods mostly involves adults, not kids. In early 2008, Google executives noticed that the word "WOW" was consistently one of the world's top ten search words. Was this a global rash of excitability? Not exactly. It's actually the abbreviation for World of Warcraft, and what people were searching for was gold. Not real gold, but virtual gold—the game's internal currency. At the time of this writing, the exchange rate was around 20 WOW gold to the U.S. dollar. Buildings in China are full of workers clicking through the game to earn these virtual assets to sell in secondary markets outside of the game.

The virtual assets trade is a good business, and in many cases it's bigger than the direct game revenues themselves. After all, why sell plastic disks at a high price once when you can sell bits over a wire for years? The people who choose to pay are, by definition, the most engaged, most committed users, and as a result the least price-sensitive and the happiest about paying. (Note that this is not unique to games: It's also the model Facebook uses with the digital "gifts" that its members can buy for each other, which account for an estimated $30 million a year for the social network.)

When you're selling disks, you risk the Hollywood "second weekend" effect: When the movie's not as good as the trailer made it look, people feel ripped off and word spreads. But in games that are free to play and only charge for items once people understand why they might want them, the risk of disappointment is lower and the odds of returning customers is higher. Simply put: You're charging the people who *want* to pay, because they understand the value of what they're getting.

"If the packaged goods games model is more like movies," says Garden, "our online games are more like TV." The aim is to build an ongoing relationship with the consumer, not just have a big weekend.

In some cases, the company running the game sells the digital items themselves. In other cases, they just create a market where players can sell virtual goods to one another, and the company makes its money from a transaction fee, like eBay does. As an example of the second model, in 2005, Sony created a marketplace, called Station Exchange, in its EverQuest II game. It let players offer in-game items for a listing fee of $1 and a 10 percent closing fee on the final price. It even offered an escrow service to ensure that people got what they paid for. It ended up being just a modest success, but it was promising enough for Sony's executives to declare it the future of the industry.

2. SUBSCRIPTIONS

In 2007, Disney announced that it was going to pay $700 million for a Web site that let children pretend to be little cartoon penguins on a patch of snow. If you think that the better part of a billion dollars is a lot to pay for a game about flightless birds, adorable or not, you probably don't have little kids. If you do, chances are you already know about Club Penguin, an online community that at the time of the purchase had attracted 12 million children (some of mine among them). In 2006 and 2007, Club Penguin mania spread through the playground with a speed normally reserved for head lice.

Club Penguin is free to play, and an estimated 90 percent of its users, who are mostly between six and twelve years old, never pay a penny for it. But if you want to "upgrade your igloo" with furniture or buy a pet for your penguin, you'll have to get your parents to whip out their credit card and subscribe for $6 a month. At the time it was purchased by Disney, Club Penguin had 700,000 paid subscribers (6 percent of all its users), who were generating more than $40 million in annual revenue.

This is one of the most common game models online, especially with those games that have a "sticky" social component. RuneScape, yet another Web-based world of orcs and elves, counts more than 1 million subscribers (out of more than 6 million users) paying $5 a month, creating a $60 million annual business. As a point of reference, that's about the same size as the subscriber user base and annual revenues of the *Wall Street Journal*'s subscription-based Web site, which is the biggest paid site of all the world's newspapers. It's also larger than the *New York Times*'s paid online subscriber base was before the paper dropped the model in favor of Free in 2008. It appears that people would rather pay to cast pretend spells than to read Pulitizer Prize–winning news. (I'll leave whether that's a good thing or a bad thing to others.)

3. ADVERTISING

In the run-up to the 2008 presidential elections, players of an Xbox Live racing game called Burnout Paradise noticed as they sped around the usual tracks that one of the billboards seemed remarkably topical. It was a picture of Barack Obama with an invitation to go to vote forchange.com, one of the campaign's Web sites. This wasn't a political statement by the game's makers, it was a paid advertisement by the Obama campaign. And it's just one of thousands of ads that now populate video games on consoles, such as the Xbox 360 and Sony PS3, and on PCs.

Some of these ads are an additional revenue stream for paid games, but an increasing number are supporting the free-to-play model. Sometimes those ads are built into the original game, but as more and more games are built to use an Internet connection it's become possible to insert ads on the fly, updating the games' billboards, posters on the walls of its cities, even the clothes that the game characters wear.

In a sense, in-game advertising has become the ultimate product placement: Not only can each player get different ads, but every time the game is played those ads can change again in an effort to ensure relevance and variety. Sometimes the ads look just like ads in the real world, and sometimes they are more subtle, from the brand of boots in a snowboarding game to the bands in its soundtrack. And sometimes the entire game is an advertisement, such as Burger King's Xbox games about racing, crashing, and sneaking around with the chain's King character.

The most common form of this is the casual games market, relatively simple games that you can play in your Web browser. The numbers here are astounding: Yahoo! Games and MTV's AddictingGames each reach more than 10 million users a month, and both are based on a free, advertising-supported game model. Overall this market is already worth more than $200 million a year, and the Yankee Group, a consultancy, estimates it will pass $700 million by 2010.

4. REAL ESTATE

Second Life is not exactly a game—it's a world where you can explore and meet other people—but it's as popular as one, with a half million active user accounts. It's free, and you can download the software and explore to your heart's content without a credit card. But if you really get into Second Life, you may want to put down roots and create your own home "in-world." For that, you'll need some land, and that is where Linden Labs, the company that runs the service, makes its money.

Linden Labs is in the virtual real estate business, and a good business it is, too. Unlike real-world realtors, Linden Labs can make as much land as it needs, and the land is made attractive by the users, who build entire towns—homes, office buildings, stores, and other attractions—themselves. Monthly lease fees range from $5 to $195, depending on the size of the plot. Or you can buy your own island for a one-time fee of $1,675, plus $295 a month.

This is not just a moneymaker for Linden Labs. It's also created a secondary market of real estate brokers within Second Life, who re-sell already developed property. One of the most successful such brokers, "Anshe Chung," claimed to have become a millionaire with such reselling.

Plenty of other online games use this model, although sometimes it's not land they're selling but space stations, castles, or even berths on pirate galleons. In a sense, this is just a subset of the class of virtual goods, like the gold of Warcraft or the clothes of Club Penguin. But the difference is that these aren't really sales—they're "land use fees," or rentals, and when you stop paying, that land or residence is typically resold to someone else.

5. MERCHANDISE

Christmas morning, 2008. Under the trees of millions of American homes, there was an otherwise ordinary stuffed animal, but for its special tag. On the tag was a code, which allowed the lucky recipient to go online and play with a virtual version of his or her own stuffed animal. This simple combination—a matched pair of plush and virtual pets—has made Webkinz the number one toy in America for two years running.

The Webkinz model is a clever combination of Free and Paid. What's the main attraction—the stuffed animal or the game? Hard to say, but it's likely that neither would have been a success without the other. In a sense it's just the natural expression of twentieth-century and twenty-first-century economics working in tandem: The atoms (the stuffed animal) cost money, but the bits (the online game) are free. While most kids have a limited appetite for stuffed animals in the real world, in the game collecting an entire menagerie is the most rewarding way to play. And the only way to add more virtual animals is to buy the stuffed ones. Thus a virtuous cycle, leading to hundreds of millions of dollars in what hasn't been a blockbuster category since the Cabbage Patch dolls.

This hybrid online/offline model is now used by everyone from Lego to Mattel, where toys come with secret codes that unlock virtual goods in the free online games on their Web sites. Another free online game for kids, Neopets, sells physical packs of trading cards of the pets, and Maple Story is doing the same for its own characters. Other games sell everything from collectible figurines to T-shirts.

It's the purest form of marginal cost economics—give away the version that costs nothing to distribute to enhance the value of the thing that has a 40 percent profit margin in stores. Free makes Paid more profitable.

FREE MUSIC

If the video game industry is a business racing toward Free to accelerate its growth, music is a business stumbling to Free to slow its decline. But the early experiments are encouraging. By now the success of Radiohead's name-your-own-price experiment with *In Rainbows* is legendary. Rather than release its seventh album into stores as usual, the band released it online with the request that you pay as much or as little as you wanted. Some chose to pay nothing, including me (not because I didn't think it was worth something, but because I wanted to see if that was, in fact, allowed), while others paid more than $20. Overall, the average price was $6.

In Rainbows became Radiohead's most commercially successful album. In an era where most music sales are falling off a cliff, Radiohead reported these jaw-dropping statistics:

- The album sold 3 million copies worldwide, including downloads from the band's site, physical CDs, a deluxe two-CD and vinyl box set, as well as sales from iTunes and other digital retailers.
- The deluxe box set, which cost $80, sold 100,000 copies.
- Radiohead made more money from the digital downloads before the release of the physical CD than the total take, across all formats, of its previous album.
- When the physical CD was released, more than two months after the name-your-own-price digital form was released, it still entered the U.S. and UK charts at number one, and the paid digital download on iTunes also entered at number one, selling 30,000 copies in its first week.
- Radiohead's tour that followed the release of the album was its biggest ever, selling 1.2 million tickets.

There are plenty of artists like Radiohead who understand the value of Free in reaching a larger audience of people who may someday become paying customers in the form of concert attendees, T-shirt buyers, or even—gasp—music buyers. Musicians ranging from Nine Inch Nails' Trent Reznor to Prince have embraced similar free distribution strategies. And there are plenty of companies outside of the core music industry that benefit hugely from free music, with Apple, whose capacious iPods would cost thousands of dollars to fill with paid music, chief among them.

But when we say the "music business," we usually mean the traditional record labels, who blame Free (mostly in the form of piracy) for their ills. That accusation may be true, but it is a mistake to equate the labels' interests with those of the music market at large. Labels traditionally package and sell recorded music, and that, as we all know, is a

HOW CAN A MUSIC CD BE FREE?

In July 2007, Prince debuted his new album, *Planet Earth*, by stuffing a copy — retail value $19 — into 2.8 million issues of the Sunday edition of London's *Daily Mail*. (The paper often includes a CD, but this was the first time it featured all-new material from a star.) How can a platinum artist give away a new release? And how can a newspaper distribute it free of charge?

Prince made money by giving away his new disk.

Prince

Potential Licensing Revenue	$5.6 M
Daily Mail Licensing Revenue	$1 M
London Concert Gross	$23.4 M
NET REVENUE	**$18.8 M**

The *Daily Mail*

Licensing Fee	$1 M
Production/Promotion	$1 M
Incremental Newsstand Revenue	$1.3 M
LOSS	**$700,000**

SOURCES: DAILY MAIL, 02 ARENA

▶ **Prince spurred ticket sales.**
Strictly speaking, the artist lost money on the deal. He charged the *Daily Mail* a licensing fee of 36 cents a disk rather than his customary $2. But he more than made up the difference in ticket sales. The Purple One sold out 21 shows at London's 02 Arena in August, bringing him record concert revenue for the region.

▶ **The *Daily Mail* boosted its brand.**
The freebie bumped up the newspaper's circulation 20 percent that day. That brought in extra revenue, but not enough to cover expenses. Still, *Daily Mail* execs consider the giveaway a success. Managing editor Stephen Miron says the gimmick worked editorially and financially: "Because we're pioneers, advertisers want to be with us."

business in terminal decline. But virtually every other part of the music market, outside the labels, is growing, often by embracing Free.

There are more bands making more music than ever before. In 2008, iTunes, the largest music retailer in the United States, added 4 million new tracks to its catalog (roughly 400,000 albums' worth!). Today, it is rare to find a band that doesn't have a MySpace page where you can listen to four songs or so for free. There are more people listening to music for more hours of the day, thanks to iPod's ability to take the music you want to hear with you everywhere. Music licensing, for television, movies, commercials, or video games, is also bigger than ever. And the mobile music industry—ringtones, "ringbacks," and the sale of individual songs—is booming. And then there's Apple itself, whose old Mac motto—"Rip, Mix, Burn"—was a winking tribute to the power of free music to sell its computers, music players, and phones.

Most of all, the concert business is thriving, driven in part by the ability of free music to enlarge the fan base. Live shows have always been one of the most profitable parts of the business. In 2002, the top thirty-five touring bands, including the Eagles and Dave Matthews Band, made four times as much from their concerts as they did from selling records and licensing, according to Allen Krueger, a Princeton economist. Some bands, such as the Rolling Stones, make more than 90 percent of their money from touring. Tickets can easily go into the hundreds of dollars, creating a thriving secondary market for resale. (In 2007 eBay bought StubHub, one of the largest such resellers.) And why not? Memorable experiences are the ultimate scarcity.

Today, the summer festival season stretches to half a year, and a generation is growing up scheduling their lives around it. And the revenues don't just come from the attendees: Tours are often sponsored (the Vans Warped Tour, for example), and companies such as Camel will pay for the right to give out free cigarettes or other products to festivalgoers. Between the food, drink, merchandise, and housing, festivals are an entire tourism business built on the lure of music that many fans never thought to pay for.

The big labels understand all too well that their role in this world is shrinking. "The music industry is growing," Edgar Bronfman, the chairman of Warner Music, told investors in 2007. "The record industry is not growing." What to do about it is another question. Some have decided to fight to keep what they have, with piracy lawsuits and often ruinous royalty demands to companies that try to create new ways for consumers to get music, such as Internet radio. Others have decided to innovate out of this mess by moving to the "360 model," where they represent all aspects of an artist's career, including touring, licensing, endorsements, and merchandise. (This has had limited success so far, mostly because the labels aren't yet very good at these other jobs, and artists often complain about the high percentage fee they charge for them.)

But some of the smaller labels are innovating more successfully, often by using Free in some form or another. RCRD LBL, a company started by star blogger Pete Rojas, offers free music supported by advertising. Name-your-price models are proliferating. And the small labels that sell the newly resurgent category of vinyl LPs to discriminating music fans routinely offer free digital downloads as tasters. In Nashville, INO Records conducted an experiment in late 2006 with a record called "Mockingbird" by Derek Webb. He explains what happened:

> I had a record I was proud of, but the label was out of marketing dollars and sales were at a trickle. So I convinced the label to let me give it away for free. But there was a catch. We were going to ask not only for names, email addresses, and zip codes for everyone who downloaded the record, we were also going to ask them to recommend the record to five friends, via email addresses they would enter (but we wouldn't keep), who they thought might want to download it too. I gave away over 80,000 records in three months. Since then I've been able to filter that email list by zip codes to find out where my fans are and then email them to get them in the rooms. I sell shows out now. And sell a lot of merch. I have a career.

There are thousands of stories like Webb's. But what's particularly interesting is that the same pragmatism about the need for the industry to embrace new models is shared even by the biggest winners of the old way. Interviewed in 2008 about the impact of file-sharing on his label, G-Unit Records, the rap artist 50 Cent had the advantage in perspective of also being an artist. Sure, file-trading was hurting his label, but there is a larger war to be won:

> The advances in technology impact everyone, and we all must adapt. What is important for the music industry to understand is that this really doesn't hurt the artists. A young fan may be just as devout and dedicated no matter if he bought it or stole it. The concerts are crowded and the industry must understand that they have to manage all the 360 degrees around an artist. They

[the industry] have to maximize their income from concerts and merchandise.

FREE BOOKS

Finally, this chapter would not be complete without a word about free books, of which this is, naturally, one (at least in digital form). Books are a special case of print, like some glossy magazines, where the physical form is still preferred by most. The book industry is not in collapse, thankfully, but that has not stopped hundreds of authors (and a few publishers) from conducting their own experiments with Free.

The big difference between books and music is that for most people, the superior version is still the one based on atoms, not bits. For all their cost disadvantages, dead trees smeared into sheets still have excellent battery life, screen resolution, and portability, to say nothing about looking lovely on shelves. But the market for digital books—audiobooks, ebooks, and Web downloads—is growing fast, mostly to satisfy demands that physical books cannot, from the need for something you can consume while driving to the need for something you can get instantaneously, wherever you are.

Most free book models are based on freemium, one way or another. Whether it's a limited-time free download of a few chapters, or the whole thing in a well-formatted PDF available forever, the digital form is a way to let the maximum number of people sample the book, in the hopes that some will buy.

For example, Neil Gaiman, the science fiction writer, gave away *American Gods* as a digital download for four weeks in 2008. The usual fears and objections were presented at first: that it would cannibalize sales in stores or, at the other extreme, that a limited availability was counterproductive since by the time many people heard about it, it would be gone. The second worry is hard to check, but the first turned out to be mistaken. Not only did *American Gods* become a best seller, but sales of *all* of Gaiman's books in independent bookstores rose by

40 percent over the period the one title was available for free. Eighty-five thousand people sampled the book online, reading an average of forty-five pages each. More than half said they didn't like the experience of reading online, but that was just an incentive to buy the easier-to-read hardcover. Gaiman then gave away his next children's book, *The Graveyard*, as free online readings in streamed video, a chapter at a time, and that, too, became a best seller.

For nonfiction books, especially those on business topics, free books are often more closely modeled after free music. The low-marginal-cost digital book is really marketing for the high-marginal-cost speech or consulting gig, just as free music is marketing for concerts. You can have the abundant, one-size-fits-all version of the author's ideas for free, but if you want those ideas tailored for your own company, industry conference, or investors meeting, you'll have to pay for the author's scarce time. (Yes, that's my model, too. Speakers Bureau details are on my Web site!)

This can even work for physical books. Consultants often buy thousands of their own volumes of strategic wisdom to distribute for free to potential clients, a tactic so common that best-seller lists now have special methods to spot and ignore these bulk sales. In Europe, newspapers sometimes offer small paperback books, sometimes in serial form, free with their issues on the newsstand, which helps drive newspaper sales. And authors increasingly offer free review copies to any blogger who wants one—the "Long Tail of book reviewers"—on the grounds that such word of mouth is well worth the few dollars each copy costs.

Like everything else in Free, this is not without controversy. Howard Hendrix, then vice president of the Science Fiction Writers of America, has called authors who give away their books "webscabs," and there are publishers who still have their doubts that free books stimulate more demand than they satisfy (sometimes based on experience). But in a world of shrinking bookstore shelf space and disappearing newspaper book review sections, authors are keen to try anything that can help them build an audience. As publisher Tim O'Reilly puts

HOW CAN A TEXTBOOK BE FREE?

College students can spend $1,000 a year on books. That's a lot, considering a $160 biology text might have a one-semester shelf life, which is why the market for used books is so big and why publishers try so hard to subvert it with tactics like new editions with different page numbers. Once this model crumbles, what will replace it? Perhaps something closer to publisher Flat World Knowledge's "open textbooks," free works that can be edited, updated, and remixed into custom course materials. But how does a publisher or author benefit from giving away a $160 textbook?

▶ **Sell more than textbooks.** A print textbook's content can be disaggregated (or versioned) into smaller chunks in a range of formats and purchasing options. The resulting menu appeals to more students who, for instance, won't read an entire book online but may purchase mp3s of a few chapters to study for midterms.

▶ **digital book (online)** .... Free
▶ **printed book (black/white)** $29.95
▶ **printed book (full color)** $59.95
(print-on-demand enables lower cost)
▶ **printable PDF – whole text** $19.95
(printing/binding would cost $40 at Kinko's)
▶ **printable PDF chapter** $1.99
▶ **audio book (mp3)** $29.95
▶ **audio chapter (mp3)** $2.99
▶ **audio summaries (10 min.)** $0.99
▶ **eBook reader – whole text** $19.95
▶ **eBook reader – chapter** $1.99
▶ **flash cards – whole text** $19.95
▶ **flash cards – one chapter** .... $.99

▶ **Entice authors.** FWK offers a better royalty rate and larger return over time. Due to a bookstore's markup, a traditional publisher nets $105 of a $160 textbook. The author gets 15 percent. In a class of 100 students, 75 will purchase the $160 text. Each subsequent semester, due to the growing availability of used copies, sales can drop by 50 percent (neither publishers nor authors profit from used book sales). By the fourth semester, five students might pay full price. Until the next edition publishes, the author's royalties and publisher's revenue continue to decline.

In the FWK model, the entry point is so much cheaper (including free) that there's little to no used book market. In a test at twenty colleges in 2008, nearly half the students paid for some form of FWK content. Though the average spent was only $30, FWK can generate the same revenue (with less overhead and lower operating costs) after six years. With a 20 percent royalty rate on all content sold, an author starts earning greater royalties in two years.

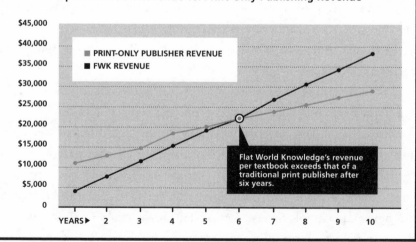

Open Textbook Revenue vs. Print-Only Publishing Revenue

- ■ PRINT-ONLY PUBLISHER REVENUE
- ■ FWK REVENUE

Flat World Knowledge's revenue per textbook exceeds that of a traditional print publisher after six years.

it, "the enemy of the author is not piracy, but obscurity." Free is the lowest-cost way to reach the largest number of people, and if the sample does its job, some will buy the "superior" version. As long as readers continue to want their books in atoms form, they'll continue to pay for them.

10

HOW BIG IS THE FREE ECONOMY?

There's More to It Than Just Dollars and Cents

I GET THIS QUESTION all the time: How big is the Free economy? To which there is only one reasonable answer: Which Free economy do you mean? It matters, because there are a lot of them, from the formal economy of business to the informal economy of volunteerism. Complicating matters further, the real ones are hard to measure, and the fake ones aren't, well, real. The countless unpaid services we do for one another every day, through kindness or social obligation, are free, but we don't tally them. And "buy one, get one free" doesn't count as a new economic model worth following.

Let's quickly dispense with the use of "free" as a marketing gimmick. That's pretty much the entire economy; I suspect that there isn't an industry that doesn't use this in one way or another, from free trials to free prizes inside. But most of that isn't really free—it's just a direct cross-subsidy of one sort or another. It's no more a distinct market than the "discount economy" might be, or any other marketing device.

Then how about the nonmonetary economies of reputation and attention? These are real economies, in the sense that they're markets with quasi-currencies that can be measured and valued, from "eye-balls" to Facebook friends. But because these are nonmonetary markets, they are, by definition, not measured in dollars and cents. Yet that hasn't stopped people from trying, often very creatively.

In early 2009, Burger King launched one of its trademark subversive marketing campaigns. Called the "Whopper Sacrifice," it offered Facebook members a free hamburger for every ten people they "unfriended" on the social network. (This was to prove that "you like your friends but love the Whopper," or more plausibly, to get some buzz-generating notoriety for Burger King.)

As it happens, there is a long tradition of measuring economies in terms of hamburgers, starting with the *Economist*'s "Big Mac Index," which compares the price of McDonald's burgers in different countries to see if their currency exchange rates are fairly valued (on the argument that a rupiah can be fiddled, but a Big Mac is a Big Mac). So bloggers quickly set out to do a similar thing with the Whopper Sacrifice and Facebook.

Facebook "friends" are a classic unit of reputational currency. The more "friends" you have, the more influence you have in the Facebook world, and the more social capital you have to spend. Indeed, most of the value of Facebook is in the fact that it has created perhaps the world's largest closed market of reputational currency, which is the foundation of its estimated multibillion-dollar valuation.

But figuring out exactly how many billions of dollars Facebook is worth has been a tricky matter. It's probably some multiple of the users it has and the number of connections between them, which is what "friending" someone creates. That act is an exchange of reputational currency, and if that currency is worth something, it must be worth something to the person giving it. But how much? And what does that imply for Facebook's valuation?

By putting a dollar value on a friend, Burger King was essentially offering a marketplace estimate of Facebook's value. Blogger Jason Kottke added it up:

Facebook has 150 million users and the average user has 100 friends. Each friendship requires the assent of both friends so really each user can, on average, only get credit for ending half of their friendships. The price of a Whopper is approximately $2.40. That means that each user's friendship is worth around 5 Whoppers, or $12. Do the math and:

$12/user × 150M users = **$1.8 billion valuation for Facebook**

Interestingly, that's considerably less than the $10 to $15 billion that the social network's investors, including Microsoft, had valued it at in 2007 and 2008. But with the economy crashing and Facebook still unable to find a way to make money faster than it is spending it, perhaps Burger King had it more right than Bill Gates. (Indeed, leaked investor documents in early 2009 showed that Facebook's internal valuation was only $3.7 billion in July 2008 and may well have fallen since.)

The value of attention and reputation is clearly something, or companies wouldn't spend so much on advertising to influence them. We set prices on attention every day: the cost to reach a thousand radio listeners for thirty seconds, the charge for forcing a million Super Bowl viewers to interrupt their game. And every time a movie star's agent negotiates a film deal, a reputation is being valued. But there's a lot more attention and reputation in the world than that measured in media and celebrity. The problem is we don't have any idea of how much more.

Is the global supply of attention fixed? Is there a given pool of attention, and for every YouTube star who ascends, another must fall to maintain some cosmic constant? Can one generation have more attention capacity than another, or does multitasking just slice the same attention capacity more finely?

Consider again the "Dunbar number," the observed limit of the number of relationships an individual can maintain in which he or she knows who each person is and how each person relates to every other person. Decades of anthropological research, studies of civilization going back millennia, fixed that number at 150. But that was before

MySpace and its kin. Now software can help you maintain links many times that. The average number of friends for MySpace members is around 180, and many go into the thousands. Has silicon enhanced our reputational capacity, or are we just diluting the meaning of "friend"?

These are all good questions, and it will probably take yet another generation to answer them. In the meantime, let's run through some of the more concrete forms of Free and get a ballpark estimate of their size.

The easiest form of Free to measure is the "three-party market," which is to say the world of advertising-supported free media we discussed earlier. Again, that's most radio and broadcast television, most Web media, and the proliferation of free print publications, from newspapers to "controlled circulation" magazines. For the top 100 U.S. media firms alone, in 2006 radio and TV (not including cable) advertising revenues were **$45 billion.**

Online, almost all media companies make their offerings free and ad-supported, as do many nonmedia companies such as Google, so I'll include the entire online ad market in the "paying for content to be free to consumers" category. That's another $21 to $25 billion. Free paper newspapers and magazines are probably a billion more. There are no doubt some other smaller categories I'm omitting and a lot of independents not included in the numbers above. Still, let's call the total of offline and online ad-driven content and services in the United States a conservative **$80 to $100 billion.**

The second form, which you're now familiar with, is freemium (what economists call "versioning"), where a few paying customers subsidize many unpaying ones. This includes both mature companies with different tiers of product pricing and start-up companies who give everything away for free while they figure out whether there will be enough demand for their offerings to lead to a business model (e.g., most Web 2.0 companies).

It's near impossible to properly tabulate all the companies who use that model, but Forrester Research, a Cambridge, Massachusetts, consultancy, has estimated that the corporate side of it (company spending on Web 2.0 services, most of which are the "premium" in the freemium equation) was around $800 million in 2008. It's a safe bet that the consumer side is at least a quarter that big, so together we can call that a round $1 billion.

Add to that the open source software market. The "Linux ecosystem" (everything from Red Hat to IBM's open source consulting business) is around $30 billion today, according to IDC, another consultancy.

WHY DO FREE BIKES THRIVE IN ONE CITY, BUT NOT ANOTHER?

In Paris, commuters can borrow a bicycle and ride for 30 minutes at no charge. Launched in 2007, ad-supported venture Vélib' (short for *vélo libre* or "free bike") now operates 1,451 stations with 20,000 bikes. Similar services are found in Barcelona, Montreal, and Washington, D.C. JCDecaux, the firm that bankrolls Vélib', also oversees flourishing programs in Lyon and Vienna. Yet, the Cyclocity bikeshare it runs in Brussels is a bust. Why do free bikes thrive in Paris, but bomb in Brussels?

A frequent commuter can spend more per year in Brussels than Paris.

Paris

€113 per year

■ ANNUAL FEE ■ USE FEE

Brussels

€264 per year

▶ **Don't nickel-and-dime riders!** In Paris, cyclists get an unlimited number of 30-minute trips with their registration fees (€1, €5 or €29 per day, week, or year). Any longer and you pay – €1 for 60 minutes, €3 for 90 minutes, €7 for two hours, etc. In Brussels, cyclists pay only €10 per year, but there's a fee for each ride – €0.50 per every 30 minutes. They got it backward: Only riders who take long trips do better in Brussels than Paris. However, the average trip in a city the size of Brussels is 20 minutes. The lesson: People prefer paying a flat fee and riding for free than feeling the shadow of a ticking meter.

▶ **More bikes at more nodes equal more users.** With 20,000 bicycles at 1,451 stations spread out across Paris, Vélib' services more residents in a variety of locales, rather than catering to specific neighborhoods. As a result, many of its riders are daily commuters. Contrast that with Brussels, where there are only 250 bicycles available at 23 stations, which are concentrated in the inner city. So why not grow the network to other parts of Brussels? Competitor Clear Channel holds contracts for certain city regions, which prevents Cyclocity from establishing its ad-driven bike hubs in those areas.

It estimates that other companies built around open source, such as MySQL ($50 million annual revenues) and SugarCRM ($15 million), probably add up to less than $1 billion.

Most of the emerging free-to-play online video game market uses the freemium model. These are primarily online massively multiplayer games, which are free to play but make money by charging the most dedicated gamers for digital assets (upgrades, clothing, new levels, etc.). The "casual games market" (think everything from online card games to flash games) is now at nearly $3 billion. Call this $4 billion total. So the total freemium market is around **$36 billion**.

Finally, there is the gift economy. This last category is impossible to properly quantify, especially since much of it has no dollar figure attached at all, but I'll give some examples that do have numbers attached so you can get a sense of scale: Apple's iPod, which gets much of its value from the fact that it can store tens of thousands of songs, only makes sense if you don't have to pay tens of thousands of dollars for that music library. Which, of course, many people don't, since they get their music free from friends or file-trading. So how much of Apple's $4 billion in annual iPod sales should be credited to Free?

Likewise, how much of MySpace's $65 billion estimated value is due to the free music bands put there? How much of the $2 billion concert business is driven by P2P file sharing? And so on. Free creates a lot of value around it, but like many things that don't travel in the monetary economy, it's hard to properly quantify. What's the value of a rainstorm or a sunny day? Both enrich the land, but the benefits are too diffuse to tabulate with any precision.

So what's the bottom line? Including both the first and second categories (ads and freemium), it's pretty easy to get to $80 billion in total revenues in the United States alone. Expand that to the traditional ad-supported media, and you can get to **$116 to $150 billion**. Go worldwide, and you can easily triple those figures, so that's at least **$300 billion** globally.

So $300 billion is a fair back-of-the-envelope guess at the Free economy, conservatively defined. It's certainly an undercount, because it doesn't consider the original form of Free—the cross-subsidy (get one thing "free," pay for another)—at all. It also doesn't do justice to the true impact of Free, which is felt as much in nonmonetary terms as it is in dollars and cents. But it does give a sense of scale: There's a lot of Free out there, and lot of money to be made off it.

A last way to calculate the size of the Free world is to look at the labor expended there. For instance, in 2008 Ohloh, a company that tracks the open source industry, listed 201,453 people currently working on a whopping 146,970 projects. That's approximately the size of the GM workforce, which is a lot of people working for free, even if not full-time. Imagine if they were producing automobiles! The author Kevin Kelly has taken this analysis to the overall Web. He notes that Google has calculated that the Web has more than 1 *trillion* unique URLs. (It's difficult to know what to count as a unique page, because a catalog can generate a different view for every visitor with every click, although Google is pretty good at distinguishing between those and hand-coded links.)

Let's say, for the sake of back-of-the-envelope calculations, that on average each page (or post or anything else with a permalink) takes one hour of research, composition, design, or programming to produce. Then the Web represents 1 trillion hours of labor.

One trillion hours over the fifteen years we've been building the Web is the equivalent of 32 million people working full-time over that period. Let's say 40 percent of that was done for free—the Facebook and My-Space pages, the blogs, the countless discussion group posts and comments. That's 13 million people—almost the working population of Canada. What would their salaries be worth, if they were paid? At a bargain rate of $20,000, that would be more than $260 billion a year.

Free is, in short, a country-sized economy, and not a little one, either.

FREECONOMICS
AND THE FREE WORLD

11

ECON 000

How a Century-old Joke Became
the Law of Digital Economics

IN 1838, ANTOINE COURNOT, a French mathematician living in Paris, published *Recherches,* now considered an economic masterpiece (although not many thought so at the time). In the book he attempted to model how companies compete, and concluded, after a lot of math, that it all had to do with the amount they produced. If one factory was making bowls and another company wanted to open a factory that also made bowls, it would be careful not to make too many, for fear of flooding the market with bowls and driving the price down. The two firms would somehow simultaneously and independently regulate their production to keep prices as high as possible.

The book was, as is often the case for even the most inspired works, promptly ignored. The members of the French Liberal School, who dominated the economics profession in France at the time, were uninterested, leaving Cournot dispirited and bitter. (He nevertheless went on to have a distinguished career, won lots of awards, and died in 1877.) But after his death, a group of younger economists returned to *Recherches* and concluded that Cournot had been unjustly neglected by his contemporaries. They called for his competition models to be reexamined.

So, in 1883, another French mathematician, Joseph Bertrand, decided to give *Recherches* a proper review. He hated it. Bertrand argued that Cournot had reached the wrong conclusion on practically everything. Indeed, Bertrand thought that Cournot's use of production volume as the key unit of competition was so arbitrary that he, half-jokingly, reworked Cournot's model with prices, not output, as the key variable. Oddly, in doing so he found a model that was just as neat, if not neater.

Bertrand concluded that rather than limit output to raise prices and increase profits, companies would more likely lower prices to gain market share. Indeed, they would attempt to undercut each other until the price was just above the cost of production, which is called "marginal cost pricing." And if the lower prices encouraged greater demand, so much the better.

Bertrand Competition can be shorthanded like this:

In a competitive market, price falls to the marginal cost.

Of course, in those days there weren't many truly competitive markets, at least not the way these mathematicians' models defined them: with homogeneous products (no product differentiation) and no collusion. So other economists dismissed the two as theoreticians trying to unnecessarily fit complex human behavior into stiff equations, and for the next few decades the spat was forgotten as yet another academic dispute.

But as economics moved into the twentieth century and markets became competitive and more measureable, researchers returned to these two feuding Frenchmen. Generations of economics graduate students labored to figure out which industries lent themselves more to Cournot Competition and which to Bertrand Competition. I'll spare you the details, but the short form is this: In abundant markets, where it's easy to make more stuff, Bertrand tends to win; price often does fall to the marginal cost.

That would still be of mostly academic interest were it not for the fact that today we are building the most competitive market the world has ever seen, one where the marginal cost of products and services is close to zero. Online, where information is a commodity and products and services can be easily copied, we are seeing Bertrand Competition playing out in a way that would have amazed even Bertrand.

If "price falls to the marginal cost" is the law, then free is not just an option, it's the inevitable endpoint. It's the force of economic gravity, and you can only fight it for so long. Yikes.

But wait. Isn't software another market with near-zero marginal costs? And doesn't Microsoft charge hundreds of dollars for Office and Windows? Yes and yes. So how does that square with the theory?

The answer lies in that part about "competitive market." Microsoft created a product that benefited hugely from network effects: The more people use a product, the more other people feel compelled to do the same. In the case of an operating system like Windows, that's because the most popular operating system will attract the most software developers to create the most programs to run on it. In the case of Office, it's because you want to exchange files with other people, so you're inclined to use the same program they use.

Both of these examples tend to produce winner-take-all markets, which is how Microsoft created a monopoly. And when you've got a monopoly, you can charge "monopoly rents," which is to say $300 for two plastic disks in a box marked "Office," when the actual cost of making those disks is just a dollar or two.

The other thing about Bertrand Competition is that it applies mostly to products that are similar. But if one product is vastly superior to another for your purposes the primary determinant of price is not marginal cost but "marginal utility"—what it's worth to you. Online, that can reflect either the features of the service or how locked into it you are.

For instance, there are many social networks out there, but if all of your own social connections are on Facebook, you may be loath to leave it, even if it were to begin charging. Its marginal utility is so much higher for you than the other social networks that you'd be willing to pay for it. But for newcomers who have not yet built their web of connections on a site, the marginal utility of the popular social networks might look more similar. Given a choice between two popular social networks—a paid Facebook and a free MySpace, say—newcomers would tend to choose the free one. And that's why Facebook doesn't charge: Its existing members might pay, but it would start losing a share of new members to free competitors.

MONOPOLIES AREN'T WHAT THEY USED TO BE

The latter half of the twentieth century was full of winner-take-all markets—jaw-droppingly high profit margins (90 percent, 95 percent, even higher) that seemed to exhibit the very opposite of Bertrand Competition. It wasn't just software, but anything where the value of the product lay mostly in its intellectual property, not its material properties. Pharmaceutical drugs (the pills cost almost nothing to produce, but the research to invent them can cost hundreds of millions of dollars), semiconductor chips (ditto), even Hollywood (movies are expensive to make but cheap to reproduce) all fall into this category.

These industries benefit from something called "increasing returns," which is to say that while the fixed costs of the product (R&D, building the factory, etc.) may be high, if the marginal costs are low, the more you make, the higher your profit margin. The rewards for pursuing a "max" strategy is that it spreads your fixed costs over a greater number of units, allowing your profit to go up with each one.

There is nothing very new about this. Even Alfred Marshall, the Victorian economist who was the first to formalize the supply and demand model, described industries where the availability of skilled labor, the presence of specialized suppliers, and the diffusion of knowledge progressively lower costs. (His prime example was cutlery makers in Sheffield, England, who were able to apply Industrial Revolution techniques to mass-produce silverware.) But "increasing returns" traditionally refers to increasing returns on production. The digital markets also benefit from increasing returns on *consumption,* where products get more valuable the more they are consumed, creating a virtuous cycle that can create market dominance.

Of course this only works if you can keep competition at bay, and the reason those profit margins were so high is that the twentieth century was full of effective ways to do that. Along with monopolies, there are patents, copyright and trademark protection, trade secrets and strong-arm tactics with retailers to keep competitors off the shelf.

The problem with most of these competition-killing strategies is that they don't work as well as they used to. Piracy, from software and content to pharmaceuticals, is growing as the technologies of duplication (from your laptop to biomedical equipment) become widespread. The largest manufacturing country in the world, China, makes pursuing patent protection difficult. And as distribution moves online, where there is infinite shelf space, it's impossible to keep competitors away from consumers, no matter how much pull you have at Wal-Mart. The Internet, by combining the democratized tools of production (computers) with democratized tools of distribution (networks), conjured the very thing that Bertrand had only imagined: a truly competitive market.

Suddenly a theoretical economic model, invented more than a century ago as a joke to ridicule another economist, became the law of pricing online.

It's too soon to say monopolies are no longer to be feared online. Those same network effects that gave Microsoft its stranglehold on the desktop work just as well on the Web, as Google has all too ably demonstrated. But what's interesting about online quasi-monopolies is that they rarely bring monopoly rents with them. For all Google's dominance, it doesn't charge $300 for its word processors and spreadsheets—it gives them away (Google Docs). Even for things it does charge for, mostly advertising space, the price is set by auction, not by Google.

So, too, for all the number ones in the big online product categories, from Facebook to eBay. For all their power, they have precious little *pricing* power. Facebook can only charge rock-bottom ad rates of less than a dollar per thousand views, and every time eBay tries to raise its listing fees its sellers threaten to leave, which, given the abundance of alternatives online, is no empty threat.

Then how do they make their billions? Scale. Not quite the old joke about losing money with each sale but making it up with volume, but instead losing money with a lot of people and making it back with a relative few. Because these companies pursue the max strategy, that relative few can still amount to thousands or millions of people. That's great news for consumers, who are getting products and services cheaply, but what about companies that can't go max? After all, for every Google and Facebook, there are hundreds of thousands of companies that never get beyond niche markets.

For them, there is no one answer: Every market is different. Free is a constant attraction across all markets, but making money around Free, especially when you don't have millions of users (and sometimes even when you do), is a matter of creative thinking and constant experimentation, of which the examples at the back of this book are just a small sample.

FREE IS JUST ANOTHER VERSION

The economic principles behind those models fall mostly into the four kinds of Free we've already discussed. And economics has no problem with prices of zero. Pricing theory is based on what's called "versioning," where different customers pay different prices. Beers at Happy Hour are cheap in the hopes that some customers will stay on and keep drinking when they're expensive.

The fundamental idea behind versioning involves selling similar products to different customers at different prices. When you decide between regular and premium gas, you're experiencing versioning, and so, too, when you see a matinee movie at half price or get a senior citizen discount. This is the core of freemium: One of the versions is free, but the others are paid. Or, to mangle Marx, to each according to her needs, from each according to her ability to open her wallet.

Another way that pricing theory can invoke Free is with flat-fee ("all you can eat") prices. You can see this in examples such as Netflix's DVD-by-mail rentals. For a fixed monthly subscription you can rent as many DVDs as you want, three at a time. Although you're still paying, you're not paying for each incremental DVD that you consume (even the postage is free). So the perceived cost of watching a DVD, sending it back, and getting a new one is effectively zero. It "feels free," even though you're paying a monthly fee for the privilege.

This is an example of what economists call near-zero "marginal price," which is not to be confused with near-zero marginal cost. The first is experienced by the consumers, the second by the producers. But the best model is when you can combine the two, which is what Netflix does.

Netflix's costs are mostly fixed: getting subscribers, keeping them, building distribution warehouses and developing software, and buying DVDs. The marginal costs of sending out more DVDs are pretty low— a little postage, some labor (although it's highly automated), and some incremental royalties—especially compared to the benefit to the subscribers of such choice and convenience. So when Netflix aligns its economic interest (spread the fixed costs over more DVDs, to lower the marginal costs) with that of its customers (flat fees make renting more DVDs feel costless), everybody wins.

In a sense, Netflix is like a gym. The fixed costs are setting up and staffing the gym. The less you use it, the more the company makes, since it can serve more members with less capacity if most of them don't show up most days. Likewise, Netflix makes more money if you don't send back your videos to be replaced often. But the difference is that you don't feel as bad about your low usage as you might with a gym. With Netflix you don't have to pay late fees if you keep a video for a few weeks, and compared to the alternative, that counts as a win.

You can see this near-zero marginal price model all around you, from the luncheon buffet to your cell phone and broadband Internet access plan. In each case, a flat fee takes the negative psychology of marginal price—the ticking meter or the feeling of being "nickeled and dimed"—off the table, making consumers more comfortable about their consumption. It works if they consume a lot, because it's usually matched with a low marginal cost production model, and it works even better (at least for the producer!) if they consume a little. As Hal Varian, Google's chief economist (and a pioneer in formalizing the economics of Free) puts it, "Who is the gym's favorite customer? It's the guy who pays his membership fee and then doesn't go."

So Free is not new to economics. It is, however, often misunderstood. One of the most famous principles that it challenges is the so-called "free-rider problem."

THE FREE-RIDER NON-PROBLEM

Russell Roberts, a George Mason economist, has a popular podcast called EconTalk (which is excellent). On one show in 2008, he observed the following:

> One of the things that fascinates me about [Wikipedia] is that I think if you'd asked an economist in 1950, 1960, 1970, 1980, 1990, even 2000, "could Wikipedia work," most of them would say no. They'd say, "Well, it can't work, you see, because you get so little glory from this. There's no profit. Everyone's gonna free ride. They'd love to read Wikipedia if it existed, but no one's going to create it because there's a free-riding problem." And those folks were wrong. They misunderstood the pure pleasure that overcomes some of that free-rider problem.

The free-rider problem is the dark side of the free lunch. Like the "free-lunch fiend" lingering in the saloon, free riders are those who consume more than their fair share of a resource, or shoulder less than a fair share of the costs of its production. But since "fair" is totally subjective, this is only considered a problem in economics when it leads to market breakdowns. So when some greedy undergrads empty the all-you-can-eat lunch buffet, causing the management to remove the buffet entirely, that would be an example of free riders running riot.

But as Timothy Lee, a computer scientist and Cato Institute scholar, has noted, the twentieth-century interpretation of this problem doesn't really work anymore, for two reasons. First, it assumes that the cost of the resource being consumed is high enough to care about or, to put it another way, that those costs must be compensated. That may be true for the lunch buffet, but it's not true for things that people happily do for free in hopes of an audience, which describes most content online. Reading them is payment enough.

And second, it grossly misjudges the effect of the Internet's scale. As we saw before, if you're the only class parent volunteer, you may eventually object to all the other parents "free riding" on your work without pitching in to help. It may upset you enough that you quit. In that case, perhaps 10 or 20 percent of the parents have to contribute to avoid the risk of the whole system breaking down.

But online, where the numbers are so much higher, most volunteer communities thrive when just 1 percent of the participants contribute. Far from being a problem, a large number of passive consumers is the reward for the few that contribute—they're called the audience.

As Lee puts it, "This large audience acts as a powerful motivator for continued contribution to the site. People like to contribute to an encyclopedia with a large readership; indeed, the enormous number of 'free riders'—aka users—is one of the most appealing things about being a Wikipedia editor."

In other words, it doesn't take a PhD to understand why Free works so well online. You just have to ignore the first ten chapters or so of your economics textbook.

The rest of this last section will look at the many sides of what's different about Free today. We'll start with the efforts to quantify nonmonetary markets such as attention and reputation, and sometimes convert them to cash. We'll then look at that paradoxical word "waste," which we're trained to avoid but should instead often pursue. (Once scarce things become abundant, markets treat them differently—exploiting the cheap commodity to create something else of more value.) Then on to China and Brazil, modern test beds of Free. And then a quick stop in fiction, where abundance as a plot device has forced authors to consider the consequences. Finally, we debate the many objections to Free, from those who question its power to those who fear it.

12

NONMONETARY ECONOMIES

Where Money Doesn't Rule, What Does?

IN 1971, at the dawning of the Information Age, the social scientist Herbert Simon wrote:

> In an information-rich world, the wealth of information means a dearth of something else: a scarcity of whatever it is that information consumes. What information consumes is rather obvious: it consumes the attention of its recipients. Hence a wealth of information creates a poverty of attention.

What Simon was observing was a manifestation of one of the oldest rules in economics: "Every abundance creates a new scarcity." We tend to value most what we don't already have in plentitude. For example, an abundance of free coffee at work awakens a need for much better coffee, for which we are willing to pay a lot. And so, too, for any premium good that arises from a sea of inexpensive commodity products, from artisanal food to designer water.

"It is quite true that man lives by bread alone—when there is little bread," observed Abraham Maslow in his groundbreaking 1943 article,

"A Theory of Human Motivation." "But what happens to man's desires when there is plenty of bread and when his belly is chronically filled?"

His answer, expressed in his now famous "pyramid of needs," was this: "At once other (and higher) needs emerge, and these, rather than physiological hungers, dominate the organism." At the base of his pyramid are physical needs, such as food and water. Above that is safety. The next higher level is love and belonging, then esteem, and finally, at the top, is "self actualization," with pursuits of meaning such as creativity.

The same sort of pyramid can be applied to information. Once our hunger for basic knowledge and entertainment is satisfied, we become more discriminating about exactly what knowledge and entertainment we want, and in the process learn more about ourselves and what drives us. This ultimately turns many of us from passive consumers to active producers, motivated by the psychic rewards of creating.

Normally in the consumer marketplace, our scarcity of money helps us navigate the abundance of products available to us—we only buy what we can afford (credit card balances notwithstanding). This is also how capitalism "keeps score" of consumer demand, by what consumers are willing to pay for. But what happens online, where more and more products are encoded into software and thus can be offered for free? No longer is money the most important signal in the marketplace. Instead, two nonmonetary factors rise in its place.

These two are what are often called the "attention economy" and the "reputation economy." Of course there is nothing new about marketplaces of attention and reputation. Every TV show has to compete in the first and every brand has to compete in the second. A celebrity builds reputation and converts it into attention. But what's unique about the online experience is how measurable the two are, and how they are becoming more like a real economy every day.

What defines an "economy"? Until the mid-1700s, the word "economy" was mostly used in politics and law. But Adam Smith gave the term its modern meaning when he defined economics as the study of markets, in particular what we now shorthand as "the science of choice under scarcity."

Today, economics studies more than just monetary markets. Since the 1970s such subspecialties as behavioral economics and "neuroeconomics" have emerged, all attempting to explain why people make the choices they do based on the incentives they experience. Attention and reputation are often part of these even if they're not formally defined as a market.

There have been some clever attempts to use the language of economics to describe attention markets, such as this nifty pirouette from Georg Franck, a German economist, in 1999:

> If the attention I pay to others is valued in proportion to the amount of attention earned by me, then an accounting system is set in motion which quotes something like the social share prices of individual attention.
>
> It is in this secondary market that social ambition thrives. It is this stock exchange of attentive capital that gives precise meaning to the expression "vanity fair."

When it came time to quantify attention back then, however, all Franck could measure was "a person's presence in the media," whatever that means.

But what if we could treat attention and reputation as quantitatively as we do money? What if we could formalize them into proper markets so we could explain and predict them with many of the same equations that economists use in traditional monetary economics? To do so, we'd need attention and reputation to exhibit the same characteristics of other traditional currencies: to be measurable, finite, and convertible.

We're actually coming close, thanks to the 1989 creation of Tim Berners-Lee: the modern hyperlink. It's a simple thing—just a string of characters starting with "http://"—but what it created was a formal language for the exchange of attention and reputation, and currencies for both. Today when you link to someone on your blog, you are effectively granting them some of your own reputation. In a sense, you are saying to your own audience: "Leave me. Go to this other place. I think you'll like it, and if you do, perhaps you'll think more of me for having recommended it. And if you think more of me, perhaps you'll come back to my site more often."

Ideally, this transfer of reputation leaves both parties richer. Good recommendations build trust with a readership, and being recommended confers trust, too. And with trust comes traffic.

Now we have a real marketplace of reputation—it's Google. What is the currency of reputation online other than Google's PageRank algorithm, which measures the incoming links that define the network of opinion that is the Web? And what better measure of attention than Web traffic?

PageRank is a deceptively simple idea with great power. It basically states that incoming links are like votes, and that incoming links from sites which themselves have lots of incoming links count for more than those that don't. This is the sort of calculation only a computer can do, since it requires having the entire link structure of the Web in memory and recursively analyzing each link. (Interestingly, PageRank is based on earlier work on a much smaller scale in scientific publishing. An author's reputation can be calculated by how many other authors cite him or her in their footnotes, a process called citation analysis. There is no more explicit reputation economy than academic reputation, which dictates everything from tenure to grants.)

In economic terms, we convert from the reputation economy to the attention economy to cash by using this formula: The economic value of your site is the traffic your PageRank (a number between one and nine) brings from Google's search results for any given term, times the keyword value for that term. (Higher PageRank means more traffic, since you'll appear earlier in the search results.) And you can convert that traffic into plain old cash by simply running AdSense ads on your site and splitting the revenues with Google.

Like it or not, we all live in the Google economy these days in at least some of our life. On a typical site, between a quarter and a half of all traffic comes from Google searches. An entire industry, called "search engine optimization," exists to help sites increase their visibility in the eyes of Google. PageRank is the gold standard of reputation.

That makes Google cofounder Larry Page (the punning Page in PageRank) the central banker of the Google economy. He and his Google colleagues control the money supply. They tweak the algorithm constantly to ensure that it retains its value. As the Web grows, they avoid PageRank "inflation" by making it harder to earn. If they see PageRank counterfeiting, in the form of linkspam, they adjust the algorithm to take it out of circulation. They maintain the value of their currency by working to keep their search results more relevant than their competitors', which will maintain Google's market share (currently a dominant 70 percent). Alan Greenspan's job was not so different.

But just as with real central bankers these days, controlling one currency is far from controlling the entire economy. Think of Google as the United States of the Web—only the biggest of many reputation and attention economies. It's not a closed economy, since it's just part of the bigger Web economy. And around it are countless other reputation and attention economies, each with its own currencies.

Facebook and MySpace have "friends." EBay has seller and buyer ratings. Twitter has "followers," Slashdot has "karma," and so on. In each case, people can build reputational capital and turn it into attention. It is up to each to figure out how to convert that to money, if that's what he or she wants (most don't), but the quantification of attention and reputation is now a global endeavor. It is a market we all now play in, whether we know it or not. Reputation that was once intangible is now increasingly concrete.

On the Web, all these economies coexist and ebb and flow with the tides of attention—even if they wanted to control attention entirely, they can't. But there is a growing class of closed online economies where the central bankers have far more power. These are online games, from Warhammer to Lineage, which typically use two currencies: an attention currency, where players earn virtual money with their game-

HOW CAN A UNIVERSITY EDUCATION BE FREE?

You don't have to enroll at UC Berkeley to watch Richard A. Muller deliver his popular "Physics for Future Presidents" lectures. They're on YouTube, along with talks from more than a hundred other Berkeley professors that have been collectively watched more than 2 million times. And Berkeley is not alone: Stanford and MIT also release lectures on YouTube, and MIT's "OpenCourseWare" initiative has put virtually all of the university's class curriculum online, from lecture notes to assignments and demonstration videos. It can cost $35,000 a year to attend these universities and take these classes. Why are they giving them away?

▶ **Lectures aren't a university education.**
Aside from the small matter of a degree, which you can't get via YouTube, a college education is more than lectures and readings. Tuition buys direct proximity to ask questions, share ideas, and solicit feedback from academics like Muller. It's access to the network of other students and the idea exchange, help, and relationships this provides. For universities, free content is marketing. Top students get their pick of schools. Sampling the mind-blowing fare of a particular program or professor can win them over.

▶ **Create demand for expertise.**
To date, one of Muller's lectures has garnered 200,000 views. That's three times the capacity of the football stadium at UC Berkeley. After becoming a Web celeb of sorts, Muller secured a book deal to write a popular hardback version of the textbook he penned for his class. Released in the summer of 2008, *Physics for Future Presidents* was widely reviewed in the mainstream press. Months later, it remained atop one of Amazon's best-seller lists. It's easy to see just how good Free has been to Professor Muller.

play, and real money, which they can use to buy virtual money if they don't want to take the time to earn it.

In each of these games, the companies behind them take their role of central banker seriously. If the Warhammer developers don't keep a cap on the gold supply, its value will fall and the resale market will collapse. Game designers often bring in academic economists to help design their in-game economies, to avoid all the ills of real-world economies, from insufficient liquidity to fraud.

But in the end, all these games pivot around the ultimate scarcity: time. Time really is money, and at the core of these game economies there is a trade-off between them. Younger players may have more time than money, and they can accumulate attention currencies with their clicks. Older players may have more money than time, and they can buy shortcuts. Game designers try to get the balance of those two right, so players can compete and advance either way. And as designers do so, they are creating some of the most quantified nonmonetary economies the world has ever seen.

THE GIFT ECONOMY

In 1983, sociologist Lewis Hyde wrote *The Gift,* one of the first books to try to explain the mechanics of one of the oldest social traditions: giving people things without charge. He focused mostly on Pacific Island and other "native" societies that had not adopted formal monetary economies. Instead, stature was established through gift exchanges and rituals—cultural currencies substituted for money.

Many of these societies lived amid abundant natural resources—food really did grow on trees—so their basic substance needs were provided by nature. Because of this, they could move up Maslow's pyramid and focus on social needs. Gifts played the role of social cement: In the case of some Native American tribes, the implicit rule of a gift was that it carried with it an obligation to reciprocate ("return the gift"). Gifts should also not be kept but instead regifted to others ("the gift must always move"). Today, we think of the term "Indian giver" as a pejorative, but it stems from what Hyde observed: In those cultures, one could never really own a gift. Instead, it was a symbol of goodwill, and only retained that if it remained in circulation.

Hyde focused mostly on gift economies of things—actual objects exchanged (as we're seeing today with Freecycle—see the sidebar on page 188). But there has always been a much larger gift economy of deeds, the things we do for each other without charge. As with the attention and reputation economies, this ephemeral gift economy has suddenly become explicit and measureable as it moves online.

In the traditional media business, where I work, you've got to pay people to write. A buck a word is the bottom of the range. Really good writers can get three bucks a word or more. If I was writing this sentence for a glossy magazine (and flattered myself by charging top dollar), I would have just earned $23. But something's changed. At last count there were 12 million active blogs, people or groups of people writing at least once a week, generating billions of words. No more than a few thousand of these writers are paid to do it.

You can see this everywhere, from Amazon's amateur product reviewers to the film buffs who have made IMDB the most comprehensive compendium of film and filmmaker information in the world. Some of it is the informal posts in the support groups of countless discussion forums, but it can also include projects that took weeks or months of work, such as player-created video game guides and catalogs of everything (there are a lot of "completionists" out there who love becoming the world's foremost expert on *something*, and sharing it with all).

There is nothing new about this—people have always been creating and contributing for free. We didn't call what they did "work" because it wasn't paid, but every time you give someone free advice or volunteer for something, you're doing something that in a different context could be somebody's job. Now the professionals and amateurs are suddenly in the same marketplace of attention, and these parallel worlds are now in competition. And there are a lot more amateurs than professionals.

What motivates the amateur creatives, if not money? Many people assume that the gift economy is driven mostly by generosity, but as Hyde observed in Pacific Islanders, it's usually not quite so altruistic. Adam Smith got it right: Enlightened self-interest is the most powerful force in humanity. People do things for free mostly for their own reasons: for fun, because they have something to say, because they want people to pay attention to them, because they want their own views to gain currency, and countless other very personal reasons.

In 2007, Andy Oram, an editor at O'Reilly Media, looked out at the amazing variety of user-generated documentation—instruction manuals for software, hardware, and games, that go beyond what the original creators provided—and wondered what motivated people to do it. He ran a survey for a year and then tabulated the results. The top reason was "community"—people felt part of a community and wanted to contribute to its vitality. The second was "personal growth," which har-

HOW CAN MILLIONS OF SECONDHAND GOODS BE FREE?

It all started with a bed. In the spring of 2003, Deron Beal discovered charitable organizations in his hometown of Tucson, Arizona, wouldn't accept his old mattress due to health concerns. To promote waste reduction, he founded Freecycle.org, a site that connects people to the stuff someone else doesn't want to take the time to sell or haul to the dump.

A nonprofit, Freecycle operates on a modest annual budget ($140,000) with only the faintest advertising (a Google sponsor bar). Driven by self-organizing Yahoo Groups run by local, volunteer moderators, Freecycle only admits users who explain (in 200 characters or less) their motives. For those who understand the cause's implicit "give and take" ethos, a plethora of free stuff awaits: leather sofas, TVs, exercise bikes, you name it.

Gift economies certainly predate the Web. But there's never been a more ef-fective platform for widespread giving. In a sense, the zero-cost distribution online has transformed sharing into an industry. Similar sites have launched: sharingisgiving.org, freecycleamerica.org, freesharing.org. On Craigslist users also post free items. However, no other site has built as active and fervent a community that relies entirely on Free.

Today, Beal's creation measures its successes not in dollars, but in the tonnage of all goods given away (600 per day!), people (5.9 million across 4,619 Yahoo Groups), and reach (85 countries). In 2008, those 5.9 million members donated roughly 20,000 items per day, nearly 8 million total – an average of at least one item per person. If each freebie could have fetched an average of, say, $50 on Craigslist, then based on current membership, the size of the Freecycle economy would be in the neighborhood of $380 million per year.

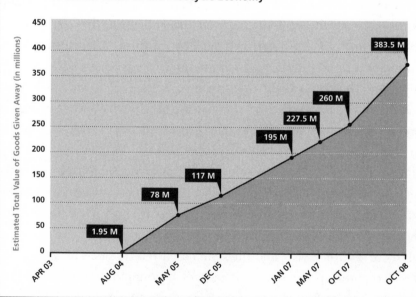

Potential Value of the Freecycle Economy

Estimated Total Value of Goods Given Away (in millions)

- 1.95 M (APR 03)
- 78 M (AUG 04)
- 117 M (MAY 05)
- 195 M (DEC 05)
- 227.5 M (JAN 07)
- 260 M (MAY 07)
- (OCT 07)
- 383.5 M (OCT 08)

kens back to Maslow's highest level, self-actualization. Third came "mutual support," which suggests that many such contributors are what sociologists call "mavens"—people with knowledge who enjoy sharing it. (Interestingly, reputation figured relatively low among the motivations in Oram's survey.)

And where do people find the time? By not doing something else—abandoning things that don't return the same social and emotional rewards. Imagine if we could harness just a fraction of the human potential lost watching TV. (Actually, there's no need to imagine that: Rating trends suggest that TV watching has already peaked, and we're increasingly choosing the screens that allow us to both produce *and* consume.)

In a world where food, shelter, and the rest of Maslow's subsistence needs are met without having to labor in the fields from dawn to dusk, we find ourselves with "spare cycles," or what sociologists call "cognitive surplus"—energy and knowledge not fully tapped by our jobs. At the same time we have emotional and intellectual needs that aren't fully satisfied at work, either. What our "free labor" in an area that we value grants us is respect, attention, expression, and an audience.

In short, doing things we like without pay often makes us happier than the work we do for a salary. You still have to eat, but as Maslow showed, there is more to life than that. The opportunity to contribute in a way that is both creative and appreciated is exactly the sort of fulfillment that Maslow privileged above all other aspirations, and what many jobs so seldom provide. No wonder the Web exploded, driven by volunteer labor—it made people happy to be creative, to contribute, to have an impact, and to be recognized as expert in something. The potential for such a nonmonetary production economy has been in our society for centuries, waiting for the social systems and tools to emerge to fully realize it. The Web provided those tools, and suddenly a market of free exchange arose.

13

WASTE IS (SOMETIMES) GOOD

The Best Way to Exploit Abundance
Is to Relinquish Control

EVERY NOW AND THEN at work I get an email from the IT department telling us that it's time for employees to "delete unneeded files from the shared folders," which is the IT way of saying that they've run out of storage capacity on their computers. Because we're good corporate citizens, we all diligently look at our folders on the server and scan through the files to see if we really need them, deleting those we can do without. Perhaps you've done the same.

One day, after years of doing this, I started to wonder just how much storage the IT department actually had for our office. To give you a sense of perspective on the answer, a terabyte of storage (1,000 GB) cost about $130 at the time I asked. Recently, when we got a standard Dell desktop PC at home, which my children use to play games, it came with a terabyte hard drive built in.

So how much storage did we have for my whole office? Turns out, not so much: 500 GB—one-half terabyte. My children had twice as much storage as my entire workplace.

How did this happen? The answer is simple: Somehow we got stuck thinking that storage was expensive when in fact it had become dirt cheap. We treated the abundant thing—hard drive capacity—as if it were scarce, and the scarce thing—people's time—as if it were abundant. We got the equation backward. (Let me hasten to add that my office quickly added a heap of storage and those emails don't go out anymore!)

This happens all over the place. When your phone company tells you that your voice mail box is full, that's artificial scarcity—it costs less than a nickel to store one hundred voice messages, and the average iPod could store thirty thousand of them (voice messages are recorded at lower quality than music, so they take less space). By forcing subscribers to take the time to delete voice mails, the phone companies were saving a little money in storage costs by spending a lot of consumer time. They managed the scarcity they could measure (storage) but neglected to manage the much larger scarcity of their customers' goodwill. No wonder phone companies are second only to cable TV companies in the "most hated" rankings.

This is a lesson about embracing waste. Just as Carver Mead preached the sermon of wasting transistors and Alan Kay responded by wasting them on the eye candy that made computers easier to use, so today's innovators are the ones who spot the new abundances and figure out how to squander them. In a good way!

But the funny thing about waste is that it's all relative to your sense of scarcity. Our grandparents grew up in an age when a long-distance telephone call was an expensive luxury, to be scheduled and kept short. Even today many people find it hard to keep people of that generation on a long-distance call for long—they still hear a meter ticking in their head and rush to finish. But our kids are growing up in an age where long-distance is free on their cell phones. They'll happily chat for hours. From the perspective of 1950s telecommunications costs, that's incredibly wasteful. But today, when those costs have fallen to near zero, we don't give it a second thought. It doesn't feel like waste. In other words, one generation's scarcity is another's abundance.

NATURE WASTES LIFE

Our brains seem wired to resist waste, but as mammals we are relatively unique in nature for this. Mammals have the fewest offspring in the animal kingdom, and as a result we invest huge time and care in protecting each one so that it can reach adulthood. The death of a single human is a tragedy, one that survivors sometimes never recover from, and we price the individual life above all.

As a result, we have a very developed sense of the morality of waste. We feel bad about the unloved toy or the uneaten food. Sometimes this is for good reason, because we understand the greater social cost of profligacy, but often it's just because our mammalian brains are programmed that way.

However, the rest of nature doesn't work like that. A bluefin tuna can release as many as 10 million fertilized eggs in a spawning season. Perhaps ten will make it to adulthood. A million die for every one that survives.

Nature wastes life in search of better life. It mutates DNA, creating failure after failure, in the hopes that every now and then a new sequence will outcompete those that came before, and the species will evolve. Nature tests its creations by killing most of them quickly, the battle "red in tooth and claw" that determines reproductive advantage.

The reason nature is so wasteful is that scattershot strategies are the best way to do what mathematicians call "fully exploring the potential space." Imagine a desert landscape with two pools of water separated by some distance. If you're a plant growing next to one of those pools, you can have one of two different reproductive strategies. You can drop seeds near your roots, where there's a pretty good chance water can be found. This is safe, but soon leads to crowding. Or you can toss the seeds into the air and let them float far away. This means that almost all will die, but it's the only way to find that second pool of water, where life can expand into a new niche, perhaps a richer one. The way to get from what the mathematicians call a "local maxima" to the "global maxima" is to explore a lot of fruitless "minima" along the way. It's wasteful, but it can pay off in the end.

Cory Doctorow, the science fiction writer, calls this "thinking like a dandelion." He writes:

> The disposition of each—or even most—of the seeds isn't the important thing, from a dandelion's point of view. The important thing is that every spring, every crack in every pavement is filled with dandelions. The dandelion doesn't want to nurse a single precious copy of itself in the hopes that it will leave the nest and carefully navigate its way to the optimum growing environment, there to perpetuate the line. The dandelion just wants to be sure that every single opportunity for reproduction is exploited!

This is how to embrace waste. Seeds are too cheap to meter. It feels wrong, even alien, to throw so much away, but it's the right way to properly take advantage of abundance.

Just consider the Roomba robotic vacuum cleaner. It's difficult to watch it and not feel sorry for its stupidity, as it bounces haphazardly around the room, retracing its steps and missing obvious patches of dirt. But eventually, somehow, the carpet gets clean as this random walk eventually covers every square inch. It may take an hour to do what you could do in five minutes, but it's not your time, it's the machine's. And the machine has plenty of time.

MAKING THE WORLD SAFE FOR CAT VIDEOS

Perhaps the best example of a glorious embrace of waste is YouTube. I often hear people complain that YouTube is no threat to television because it's "full of crap," which is, I suppose, true. The problem is that none of us can agree on what "crap" is, because we can't agree on its opposite, "quality." You may be looking for funny cat videos, and my favorite soldering tutorials are of no interest. I, meanwhile, want to see funny video game stunts, and your cooking tutorial is of no interest. And videos of our own charming family members are of course delightful to us and totally boring to everyone else. Crap is in the eye of the beholder.

Even the most popular YouTube videos may totally fail the standard Hollywood definition of production quality, in that the videos are low-resolution and badly lit, their sound quality awful and their plots nonexistent. But none of that matters, because the most important thing is relevance. We'll always choose a "low-quality" video of something we actually want over a "high-quality" video of something we don't.

A few weekends ago it was time for my kids to choose how to spend the two hours of "screen time" they're allowed on Saturdays and Sundays. I suggested that it was a great day for *Star Wars* and gave them a choice. They could watch any of the six movies on magnificent DVD, on a huge hi-def projection screen with surround sound audio and *popcorn*. Or they could go on YouTube and watch Lego stop-motion animations of *Star Wars* scenes created by nine-year-olds. It was no contest—they raced for the computer.

It turns out that my kids, and many like them, aren't really that interested in *Star Wars* as created by George Lucas. They're more interested in *Star Wars* as created by their peers, never mind the shaky cameras and fingers in the frame. When I was growing up, there were many clever products designed to extend the *Star Wars* franchise to kids, from toys to lunch boxes, but as far as I know nobody thought of Lego stop-motion animation created by children.

The demand for stop-action *Star Wars* must have always been there, but just invisible because no marketer thought to offer it. But once we had YouTube, and didn't need a marketer's permission to do things, an invisible market suddenly emerged. Collectively, we found a category that the marketers had missed. (There are dozens of other amateur *Star Wars* markets like this, from fan fiction to the 501st Legion of grown-ups who make their own amazing storm trooper suits and gather for reenactments.)

All those random videos on YouTube are just dandelion seeds in search of fertile ground on which to land. In a sense, we're "wasting video" in search of better video, exploring the potential space of what the moving picture can be. YouTube is a vast collective experiment to invent the future of television, one thoughtless, wasteful upload at a time. Sooner or later, through YouTube and others like it, every video that can be made will be made, and every filmmaker that can be a filmmaker will become one. Every possible niche will be explored. If you lower the costs of exploring a space, you can be more indiscriminate in how you do it.

Nobody is deciding whether a video is good enough to justify the scarce channel space it takes, because there is no scarce channel space. Distribution is now close enough to free to round down. Today, it costs about $0.25 to stream one hour of video to one person. Next year it will be $0.15. A year later it will be less than a dime. Which is why YouTube's founders decided to give it away, both *gratis* and *libre*. The result is messy and runs counter to every instinct of a television professional, but this is what abundance both requires and demands. If YouTube hadn't done it, someone else would have.

What this boils down to is the difference between abundance and scarcity thinking. If you're controlling scarce resources (the prime-time broadcast schedule, say) you have to be discriminating. There are real costs associated with those half-hour chunks of network time, and the penalty for failing to reach tens of millions of viewers with them is calculated in red ink and lost careers. No wonder network executives fall back on sitcom formulas and celebrities—they're a safe bet in an expensive game.

But if you're tapping into abundant resources, you can afford to take chances, since the cost of failure is so low. Nobody gets fired when your YouTube video is only seen by your mom.

For all YouTube's successes, however, it has so far failed to make any money for Google. The company has not figured out how to match video ads with content, the way it matches text ads with text content on the Web. It doesn't really know what the video you uploaded is about, and even if it did, it probably doesn't have a relevant video ad to match with it. Meanwhile, advertisers are distinctly uncomfortable with their brands being placed against user-generated content, which can be offensive.

The TV networks saw an opportunity in this failing and created a competing video service, Hulu. It offers mostly commercial video, most of it taken from TV, but is as convenient and accessible as YouTube. Because this content is a known quantity, often the same thing that the advertisers are already buying on TV, they're happy to insert their commercials as pre-rolls, post-rolls, and even interruptions in the programming. It's free, of course, but unlike YouTube, you're paying with your time and annoyance—just like regular TV. But if it's *30 Rock* you want, and you want it now, in your browser, this is the only way you're going to get it.

SCARCITY MANAGEMENT

The YouTube model is totally free—free to watch, free to upload your own video, free of interruptions. But it doesn't make money. Hulu is only free to watch, and you have to pay the good old-fashioned way, by watching ads you may or may not care about. Yet it generates healthy revenues. The two video outlets illustrate the tension between different models of Free. Although consumers may prefer 100 percent free, a little artificial scarcity is the best way to make money.

I see this every day as a magazine editor, where I live in both worlds. In print, I operate by the rules of scarcity, since each page is expensive and I've got a limited number of them. Since saying yes to a story proposal is so costly, from the dozens of people who will be involved to the factories that may someday print the words on the page, my job is to say no to almost everything. Either that's explicitly rejecting proposals or, more typically, setting the bar so high that most proposals don't get to me in the first place. Because I'm responsible for allocating costly resources, I fall back on a traditional top-down management hierarchy, with a chain of approvals necessary to get something into print.

Not only are our pages expensive, they are also unchangeable. Once the presses run, our mistakes and errors of judgment are preserved for eternity (or at least until they're recycled). When I make a decision in the production process, we are committed to a path that it is expensive to deviate from. If something better comes along, or my decision doesn't look as smart as it did a few weeks earlier, we sometimes have to continue anyway, making the best of it. In this case, we are forced to focus on economic costs, ignoring the potentially even larger *opportunity costs* of all the paths not taken because of our scarcity-driven publishing model.

Online, however, pages are infinite and infinitely changeable. It's an abundance economy and invites a totally different management approach. On our Web site we have dozens of bloggers, many of them amateurs, who write what they want, without editing. On parts of the site we invite users to contribute their own content. Our default response to story ideas can be yes or, more to the point, "Why are you even asking me?" The cost of a dull story is mostly that it won't be read, not that it will displace a potentially more interesting one. Successes rise to the top, while failures fall to the bottom. Everything can get out there and compete for attention, winning or losing on its merits, not a manager's guesswork about what people want.

The reality of managing these two worlds is not quite so black-and-white, of course. Even though we have unlimited pages online, we still have a reputation to keep up and a brand to preserve, so it's no free-for-all. Instead, it's a hybrid structure, where costs and control tend to move in parallel; the lower the costs, the less control we have to exercise. Standards such as accuracy and fairness apply across the board, but in print we have to try to get everything right before publishing, at great expense, while online we can correct as we go. Because we compete in both scarce and abundant markets, one-size management structure doesn't fit all—we need to simultaneously pursue both control and chaos.

Sound schizophrenic? It's just the nature of the hybrid world we're entering, where scarcity and abundance exist side by side. We're good at scarcity thinking—it's the twentieth-century organizational model. Now we have to get good at abundance thinking, too. Here are some examples of how that works:

	Scarcity	Abundance
Rules	"Everything is forbidden unless it is permitted"	"Everything is permitted unless it is forbidden"
Social model	Paternalism ("we know what's best")	Egalitarianism ("you know what's best")
Profit plan	Business model	We'll figure it out
Decision process	Top-down	Bottom-up
Management style	Command and control	Out of control

14

FREE WORLD

China and Brazil Are the Frontiers of Free. What Can We Learn from Them?

I'M IN A huge banquet hall in Guangzhou, China, sitting in the front row of a spectacular show. We've already had the acrobatics, the kung fu exhibition, the dancing girls, and the comedy act. Now it's time for the real star, Taiwanese pop sensation Jolin Tsai. The audience cheers as she sings some of her best-known numbers, her gown shimmering in the lights against a backdrop of her face blown up to room size on a huge video screen.

This is not a concert, however. It's a sales meeting of China Mobile employees and partners. We've had a day of speeches on the telecom business, and it is just customary to finish off with a fancy show. For her performance, Tsai was probably paid more than she earned from CD sales all year.

China is a country where piracy has won. Years of halfhearted crackdowns under diplomatic pressure from the West have had no apparent effect on the street vendors or countless sites that host MP3s for downloads. Every year there are some ceremonial piracy busts, and some of the bigger Web sites occasionally have to pay fines, but none of this has stopped average Chinese music consumers from finding pretty much everything they want for free.

So rather than fight piracy, a new breed of Chinese musician is embracing it. Piracy is a form of zero-cost marketing, which brings their work to the largest possible audience. That maximizes their celebrity (at least for the brief duration of any Chinese pop star's fame), and it is up to them to find ways to convert that celebrity into cash.

Xiang Xiang is a twenty-one-year-old Chinese pop star, most famous for her cheeky song "Song of Pig." Her latest album sold nearly 4 million copies. The problem is that almost all of them were pirated versions. Or rather, that's her label's problem. She's fine with it. As far as she's concerned, that's 4 million fans she wouldn't have had if they'd had to pay full price for the album, and she likes the feedback and adulation. She also likes the money she gets for personal appearances and product endorsements, all made possible by her piracy-enabled fame. And then there's the concert tour, which should take her to fourteen cities this summer. The pirates are her best marketers.

Piracy accounts for an estimated 95 percent of music consumption in China, which has forced record companies to completely rethink what business they're in. Since they can't make money from selling music on plastic disks, they package it in other ways. They ask artists to record singles for radio play instead of albums for consumers. They serve as a personal talent agency for the singers, getting a cut of their fees for making commercials and radio spots. And even concerts are paid for by advertisers brokered by the labels, which pack as many of their artists on stage as they can to maximize the revenues from sponsors. The main problem is that the singers complain that the endless touring, which provides their only income, is tough on their vocal cords.

"China will become a model for the world music industry," predicts Shen Lihui, who runs Modern Sky, one of the more innovative Chinese music labels. The company's CDs rarely make money because the popular ones are quickly pirated. But the label has other ways of making money: producing videos and now, increasingly, Web sites. It also runs a three-day music festival that attracts fans from around the country. Ticket sales are part of the revenues, but corporate sponsors are where the real money is: Motorola, Levi's, Diesel, and others.

That's not to say that you can't sell music in China: You can, as long as the songs are less than twenty seconds long. The ringtone and ringback business is huge: China Mobile, the largest carrier, reported more than a billion dollars in music revenues in 2007. Most of that was kept by China Mobile, of course, but that's still real money.

Ed Peto, a Briton living in Beijing, is trying to find another way to turn music into a business. His company, MicroMu, signs emerging indie artists and gets brands to sponsor the entire operation with a monthly fee. The way it works may sound strange to a Western record label, but it makes perfect sense to a product marketer in China, who is actually the paying customer in the equation.

MicroMu records artists as cheaply as possible, either as a live recording in front of an audience at a sponsored show or in inexpensive studio space or a rehearsal room. They film everything surrounding these sessions and make a range of branded video content from this. Each recording is released through a blog post on the MicroMu site, complete with links to free downloads of individual MP3s, full album downloads, credits, artwork, etc. Then the company hosts regular live events, including university tours.

Brands such as jeans and drink companies sponsor MicroMu, but not the individual artists (to avoid tarnishing their indie credibility). A percentage of the sponsorship money is divided up among the artists according to how many downloads they get through the site.

"The moment you put a fee on accessing music in China is the moment you cut off 99 percent of your audience," says Peto. "Music is a luxury for the middle class in China, a flippant expenditure. This model works against that. We simply use free music and media as a way of saying that 'everyone is welcome,' building a dialogue, building a community, becoming the trusted brand of the grassroots music movement in China. To do this, though, we have to become all things to all men: record label, online community, live events producers, merchandise sellers, TV production company."

As goes China, so may go the rest of the world. U.S. record sales fell by nearly 15 percent in 2008, and the bottom is nowhere in sight.

The day may come when many labels simply capitulate and follow the Chinese model, letting music go free to become marketing for the talent, whom they monetize in nontraditional ways, such as endorsements and sponsorships. There are already some glimpses of this: The deal Madonna has with Live Nation is based on a share of all her revenues, including touring and merchandise. And talent agencies such as CAA and ICM are considering becoming music labels, to cut out the middleman. In a world where the definition of the music industry is changing every day, the one constant is that music creates celebrity. There are worse problems than the challenge of turning fame into fortune.

THE CHANEL KNOCKOFF ECONOMY

Piracy doesn't stop at film, software, and music in China. Just get off the train in Shenzhen and you are immediately bombarded with knockoff Rolex watches, Chanel perfume, and Gucci bags, and countless ersatz toys and gadgets. Like the pirated CDs on the street corners, these aren't actually free, of course, they're just very cheap; it's only that the original creators aren't seeing a penny of the sales. The intellectual property rights are free; you just pay for the commodity atoms. But as with music, the roots and consequences of this piracy are more subtle than they appear.

Piracy extends to virtually every industry in China, a combination of the state of development of the country and its legal systems and a Confucian attitude toward intellectual property that makes copying the work of others both a gesture of respect and an essential part of education. (It's often hard to explain to Chinese students in the United States what's wrong with plagiarism, since reproducing the masters is so central to Chinese teaching.) Today, an entire industry exists in China to clone designer goods overnight: Software allows factories to take photographs of Fashion Week models off the Web and produce simulations of designer clothing within a couple of months, often beating the originals to the stores.

In the Western press, Chinese piracy is seen as little more than a crime. Yet within China, pirated goods are just another product at another price, a form of market-imposed versioning. The decision whether to buy a pirated Louis Vuitton bag is not a moral one, but one about quality, social status, and risk reduction. If people have the money, they'd still rather buy the real thing, because it's usually better. But most people can only afford the pirated versions.

Just as the Cantopop download sites create celebrity while they displace sales, the pirates aren't just making money on somebody else's designs; they're also serving as a form of zero-cost brand distribution for those designs. A fake Gucci bag still says Gucci, and it's *everywhere*. This has mixed consequences: a combination of the negative "replacement effect" (the pirated versions take demand that the authentic versions might have tapped) and a positive "stimulus effect" (the pirated versions create brand awareness that can be tapped elsewhere).

In 2007, the China Market Research Group surveyed consumers, mostly young women, in big Chinese cities and found an essentially pragmatic approach to piracy. These consumers understood the difference between the original and pirated products, and preferred the originals if they could afford them. Sometimes they would buy one original and then complete their outfit with fakes.

One of the researchers, Shaun Rein, reported that some young women making $400 a month said that they were willing to save three months of salary to buy a thousand-dollar Gucci handbag or shoes from Bally. One twenty-three-year-old female respondent said, "Right now I can't afford to buy a lot of real Prada or Coach, so I buy the fake items. I hope that in the future I will be able to afford the real thing, but right now I want to look the part."

Looking the part doesn't always mean just buying convincing fakes. It has also created a market for fake *evidence* that the product *isn't* fake. (There's innovation for you.) You can buy large-sum price tags to attach to your low-sum clothes (it's not uncommon to see people wearing sunglasses with the price tag still attached), and there is even a secondary market in fake receipts. The products are one thing, but the status that goes along with them is much more important.

A twenty-seven-year-old woman working at a multinational admitted that she did buy fakes but said, "If you wear a lot of fake clothing or have a lot of counterfeit bags, your friends will know, so you are not fooling anybody. It is better to have the real thing."

This highlights the difference between digital and physical things. Pirated digital products are as good as the originals. But pirated physical products usually aren't. After years of scandals involving knockoff Chinese baby and pet food with inferior and sometimes poisonous ingredients, Chinese consumers are hyperaware of the risks associated with buying in the gray economy.

The throngs of Chinese consumers traveling to Hong Kong to buy certified luxury goods is testament to the effect of the ubiquitous pirating of designer products on the mainland: Consumers are very aware of Western luxury brands, they associate them with style and quality, and they're keen to buy the real thing when they can. And increasingly, they can.

Piracy didn't destroy the market—it *primed* the market for an emerging tide of middle-class consumers. Per capita income has more than doubled in China in the last decade, from $633 in 1996 to $1,537 in 2007, and shows few signs of slowing. There are now around 250,000 millionaires in China, and the number is growing every day. Today, China (including Hong Kong) is the third-largest market for legitimate luxury goods in the world. In economic terms, piracy stimulated more demand than it satisfied.

The idea that knockoffs can actually help the originals, especially in the fashion business, isn't new. In economics, it's called the "piracy paradox," a term coined by law professors Kal Raustiala and Christopher Sprigman.

The paradox stems from the basic dilemma that underpins the economics of fashion: Consumers have to like this year's designs, but also quickly become dissatisfied with them so they'll buy next year's design. Unlike technology, say, apparel companies can't argue that next year's models are functionally better—they just look different. So they need some other reason to get consumers to lose their infatuation with this year's model. The solution: widespread copying that turns an exclusive design into a mass-market commodity. The designer mystique is destroyed by cheap ubiquity, and discriminating consumers have to go in search of something exclusive and new.

This is what Raustiala and Sprigman call "induced obsolescence." Copying allows fashion to move quickly from early adopters to the masses, forcing the early adopters to adopt something new. In China the early adopters are the emerging rich and middle class, while the masses are a billion people who can still dip their toe in the luxury market with a clever fake. The two products—real and knockoff—are simply targeted at different market segments. Each feeds the other. And it's not just in China.

THE POWER OF BRAZILIAN STREET VENDORS

On a busy corner in São Paulo, Brazil, street vendors pitch the latest "tecnobrega" CDs, including one by a hot band called Banda Calypso. *Brega* doesn't have a direct translation, but it roughly means "cheesy" or "tacky," and the music, which comes from the poorer northern state of Pará, has an unruly party sound, traditional Brazilian music driven by a techno beat. Like CDs from most street vendors, these are not the official offerings from a big label. But neither are they illicit.

The CDs are created by local recording studios, which tend to be run by local DJs. They, in turn, get the masters from the band itself, along with CD liner art. The local DJs work with local party planners, street vendors, and radio stations to promote the upcoming show. Sometimes the local DJs are actually all of these combined, producing, selling, and promoting the CDs for the show that they themselves are organizing.

Banda Calypso doesn't mind that it doesn't get any money from this, because selling disks isn't Calypso's main source of income. The band is really in the performance business—and business is good. Moving from town to town this way, preceded by a wave of supercheap

CDs, Calypso can fill hundreds of shows a year. The band usually plays two or three shows a weekend, traveling around the country by microbus or boat.

But it's not all road and river trips. Hermano Vianna, an anthropologist and scholar of Brazilian music, tells a story about Calypso to illustrate their success. While planning a feature on the band for his Globo TV music show, Vianna offered a Globo-owned airplane to get the band to and from a show in a remote area of the country. Calypso's reply? No need, we have our own plane.

In a sense, the street vendors have become the advance team in each town Calypso visits. They get to make money from the music CDs, which they sell for as little as $0.75, and in turn they display the CDs prominently. Nobody thinks of the vendors' cheap CDs as piracy. It's just marketing, using the street economy to generate literal street cred. As a result, by the time Calypso comes to town, everyone knows about it. The band gets huge crowds to its "soundsystem" events, where it not only charges for admission, but also food and drinks. The band's crew also records the show and burns CDs and DVDs on the spot, selling them for around $2 so concertgoers can replay the show they just saw.

More than 10 million of Calypso's CDs have been sold, mostly not by the band itself. And they're not alone. The tecnobrega industry now includes hundreds of bands and thousands of shows each year. A study by Ronaldo Lemos and his colleagues at the Center for Technology and Society at Rio de Janeiro's Getulio Vargas Foundation found that between the shows and the music, this industry generates around $20 million in revenue a year.

Ninety percent of the bands have no record contract and no label. They don't need one. Letting others get their music for free creates a bigger industry than charging ever could. This is something that Brazil understands better than most: Its culture minister until 2008, the pop star Gilberto Gil, has released his music under a free Creative Commons license (including in a CD we distributed for free with *Wired*).

As in China, the drive toward Free in Brazil goes far beyond music. In 1996, in response to Brazil's alarming rate of AIDS infection, the government of then-president Fernando Henrique Cardoso guaranteed distribution of the new retroviral drug cocktails to all HIV carriers in the country. Five years later, with the AIDS rate dropping, it was clear that the plan was wise but—at the prices being charged for the patented drugs in the cocktail—utterly unsustainable.

So Brazil's health minister went to the key patent holders, the U.S. pharmaceutical giant Merck and the Swiss firm Roche, and asked for a volume discount. When the companies said no, the minister raised the stakes. Under Brazilian law, he informed them, he had the power in cases of national emergency to license local labs to produce patented drugs, royalty-free, and he would use it if necessary. The companies caved, and prices fell by more than 50 percent. Today, Brazil has one of the largest generic drug industries in the world. Not free, but royalty-free, an approach to intellectual property rights that the industry shares with the tecnobrega DJs.

Then there's open source, in which Brazil is a global leader. It built the world's first ATM network based on Linux. The prime directive of the federal Institute for Information Technology is to promote the adoption of free software throughout the government and ultimately the nation. Ministries and schools are migrating their offices to open source systems. And within the government's "digital inclusion" programs—aimed at bringing computer access to the 80 percent of Brazilians who have none—Linux is the rule.

"Every license for Office plus Windows in Brazil—a country in which 22 million people are starving—means we have to export sixty sacks of soybeans," Marcelo D'Elia Branco, coordinator of the country's Free Software Project, told writer Julian Dibbell. From his perspective, free software isn't just good for consumers, it's good for the nation.

15

IMAGINING ABUNDANCE

Thought Experiments in "Post-Scarcity" Societies, from Science Fiction to Religion

ALL SCI-FI AUTHORS know the unwritten rule: You can only break the laws of physics once or twice per story. After that, the rules of the regular world hold. So you can have time travel, invent the Matrix, or put us on Mars. But aside from that, we're just everyday folk. The fun in the stories comes from seeing how humanity responds to this one big dislocation.

Science fiction is what writer Clive Thompson calls "the last bastion of philosophical writing." It's a sort of simulation, Thompson says, where we change some of the basic rules and then learn more about ourselves. "How would love change if we lived to be five hundred? If you could travel back in time to reverse decisions, would you? What if you could confront, talk to, or kill God?"

One device authors come back to time and time again is the invention of some machine that makes scarce things abundant. It's the matter replicators in *Star Trek* (any material good you want is a button-push away) and the robot-run offworld of *WALL-E* (where people become corpulent blobs as they pass their days reclining in levitating pool loungers, drinks always at hand).

In sci-fi circles (and for the more fringy techno-utopians) this is called "post-scarcity economics." In that context, many of these novels are not just stories, they're also book-length thought experiments about the consequences of expensive things becoming close to free.

Take E. M. Forster's 1909 short story "The Machine Stops." One of the earliest examples of the post-scarcity form presents a world where humanity has retreated underground, living in individual cells, separate from all physical interaction. A colossal Machine provides for all life, ensuring food, entertainment, and protection from the toxic surface world, like some vast mechanical god. Indeed, the people in the story eventually grow to worship the Machine. And why not? Within the rooms, all human needs are taken care of:

> There were buttons and switches everywhere—buttons to call for food, for music, for clothing. There was the hot-bath button, by pressure of which a basin of (imitation) marble rose out of the floor, filled to the brim with a warm deodorized liquid. There was the cold-bath button. There was the button that produced literature. And there were of course the buttons by which [Vashti, the main character] communicated with her friends.

Vashti has no occupation or purpose other than to give lectures to her friends via their instantaneous videocommunications devices. (For the parents of today's tweens and teens, I'm sure this sounds familiar.)

There you have it: a picture of abundance. And how does it work out? Not so well. Because the Machine intermediates all personal interactions, people lose their face-to-face communication skills and become terrified of actually meeting one another. The inhabitants of the Machine decide that all information should be gathered, third-, fourth-, even tenth-hand to avoid direct experience. Unfortunately, this avoidance of interaction means an end to all collaborative creativity, and progress stops. Humanity loses its sense of purpose, giving up even the creation of art and writing to the Machine.

When the Machine eventually begins to break down, no one knows how to fix it. So as the Machine crumbles, the people of the Earth die en masse, crushed alive in the underground hives. In the end, though, one character reveals with his dying breaths that he has discovered a society of exiles still living on the surface of the Earth, free from the prison of abundance. Whew!

Other science fiction from the early twentieth century took a similarly bleak tone. The dislocations of the Industrial Revolution were still under way, and the wrenching social change brought about by mechanization, urbanization, and globalization was unsettling. Machine-created abundance was seen as available only to the privileged few—the industrialist beneficiaries of the factories in which others labored.

In Fritz Lang's *Metropolis,* society is divided into two groups: one of planners and thinkers living in luxury high above the Earth, and another of workers, dwelling and toiling underground to run the machine that sustains the wealthy. The film is about the workers' revolt, but the broader point is clear. Abundance comes at a cost: scarcity elsewhere.

The world wars put a damper on most sci-fi utopianism, but the dawn of the Space Age brought it back, and this time without as much of a dark edge. As with the stories above, Arthur C. Clarke's 1956 *The City and the Stars* starts with a hermetically sealed techno-city, where machines supply everything that's needed and nobody ever really dies. The citizens fill their days with philosophical discussions, making art and taking part in virtual reality adventures. After a few thousand years, they return to the Hall of Creation to have their consciousness digitized again. Clarke portrays this as idyllic but a bit short of meaning; his central character decides to venture out into the surrounding desert to see if there is more and eventually finds a world much like our own, where the normal cycle of birth and death provide purpose.

The arrival of the digital age and the Internet gave science fiction a more plausible source of abundance: computers. Jack yourself into the Metaverse and you could be anyone you want; scarcity was simply a construct of the virtual reality, and if you hacked it right you could have anything. Contemporary writers took a more positive view of abundance because they were already experiencing it—the Internet brought the end of information scarcity.

Of course plot requires tension, so all is not well in these utopias of abundance. In Cory Doctorow's *Down and Out in the Magic Kingdom,* some undescribed technology controlled by the Bitchun Society had "all but obsoleted the medical profession: why bother with surgery when you can grow a clone, take a backup, and refresh the new body? Some people swapped corpuses just to get rid of a cold." The result, however, is that people become bored and apathetic. One character explains: "Junkies don't miss sobriety, because they don't remember how sharp everything was, how the pain made the joy sweeter. We can't remember what it was like to work to earn our keep; to worry that there might not be enough, that we might get sick or get hit by a bus."

What becomes scarce in Doctorow's world is reputation, or "whuffie." It serves as a digital currency, something that can be given to people in exchange for good deeds, and diminished by bad behavior. Head-up displays reveal everyone's whuffie, which serves as a measure of status. When all physical needs are met, the most important commodity becomes social capital.

In Neal Stephenson's *The Diamond Age or, A Young Lady's Illustrated Primer,* the abundance comes from nanotech-driven "matter compliers," which can make anything from a mattress to food. What work remains is in designing new things for the compliers to make, and it doesn't take many people to do that. Two billion workers are idle. The book follows the efforts of one man to invent a way to educate them (thus the primer). This echoes a similar theme from writers at the time of the first Industrial Revolution: When the machines do all the work, what motivates us?

In some of these books, the end of labor scarcity liberates the mind, ends wars over resources, and creates a civilization of spiritual, philosophical beings. In others, the end of scarcity makes us lazy, decadent, stupid, and mean. You don't have to spend much time online to find examples of both.

AFTERLIFE

Perhaps no greater example of abundance/scarcity-driven extremes exists than in religion.

Heaven is the ultimate imagining of abundance: Angels drift in fluffy clouds, playing harps and transcending physical needs. Those who die holy become incorruptible, glorious, and perfect. Any physical defects the body may have labored under are erased. Islamic texts are more explicit about the specifics: Inhabitants will be of the same age (thirty-two years for men) and of the same stature. They will wear costly robes, bracelets, and perfumes and, reclining on couches inlaid with gold or precious stones, partake in exquisite banquets, served in priceless vessels by immortal youths. Foods mentioned include clear drinks neither bringing drunkenness nor rousing quarreling.

George Orwell satirized this vision of abundant paradise. In *Animal Farm,* the livestock were told that after their miserable lives were over they would go to a place where "it was Sunday seven days a week, clover was in season all the year round, and lump sugar and linseed cake grew on the hedges."

But it doesn't take many *New Yorker* cartoons to reveal that if we take the abundance myth of heaven too seriously, we quickly imagine how bored we'd be there. Robes, harps, every day like the last—blah. No wonder abundance in fiction quickly leads to total loss of purpose and the bloated sloth of *WALL-E.* Is it inevitable that the end of scarcity also means the end of discipline and drive?

It's worth looking at a historical analogy, the civilizations of Athens and Sparta, for an answer. Supported by massive populations of slaves, both of these two classical cities lived in functionally abundant worlds. The slaves provided for all the corporeal needs, much like the Machine or the Bitchun Society. If you were lucky enough to be born into the right class, you didn't have to work to live.

Neither society found itself floundering or stagnating for lack of purpose. The Athenians became artists and philosophers, trying to seek purpose in the abstract, while the Spartans focused their lives on military strength and might. Rather than depriving life of purpose, material abundance created a scarcity of meaning. Athenians moved further up Maslow's pyramid, exploring science and creativity. And the Spartans' lust for battle? I suppose Maslow would call that a form of self-actualization, too.

The lesson from fiction is that we can't really imagine plenty properly. Our brains are wired for scarcity; we are focused on the things we don't have enough of, from time to money. That's what gives us our drive. If we get what we're seeking, we tend to quickly discount it and find a new scarcity to pursue. We are motivated by what we don't have, not what we *do* have.

This is why readers under thirty, when told of the Internet's economic bounty of near-zero marginal costs, so often think "Duh." In the old paradigm, digital goods too cheap to meter counted as an almost unimaginable cornucopia. But in the new paradigm, it's hardly worth counting at all. Abundance is always the light on the *next* peak, never the one we're on. Economically, abundance is the driver of innovation and growth. But psychologically, scarcity is all that we really understand.

I'll end with an example from the dawn of the Industrial Age in Shropshire, England. In 1770, the local ironmaking factories had developed the ability to cast large sections of iron. To demonstrate the advantages of this durable new building material, the factory owners commissioned engineers to build a bridge entirely out of iron. The Iron Bridge, which spans the river Severn, remains a tourist attraction today, and it is notable not just for the ambitions of its builders and the marvel it inspired, but also because it was built entirely on timber principles.

Each element of the frame was cast separately, and fastenings followed those used in woodworking, such as the mortise and blind dovetail joints. Bolts were used to fasten the half ribs together at the crown of the arch. Thousands of these metal planks were fixed together, just as if they had been milled from a metal forest. As a result, the bridge was wildly overdesigned, and within a few years its masonry started to crack under the pressure of 380 tons of iron.

It took several decades before people realized that iron could be made to work differently. Wood's constraints of length and radial weakness were not iron's constraints. Metal bridges could have much longer arches and could be welded. Subsequent bridges the size of the Iron Bridge weighed less than half as much. People don't always recognize abundance when they first see it.

16

"YOU GET WHAT YOU PAY FOR"

And Other Doubts About Free

LATE IN 2007, Andrew Rosenthal, the editorial page editor of the *New York Times,* was interviewed by *Radar* magazine about, among other things, the *Times*'s decision to open up all its content for free online by eliminating the Times Select paywall previously maintained for its columnists.

He said the following:

I think that the newspaper industry joined hands and took a collective leap off a cliff for no discernible reason when we decided to announce to the world that what we do has no value at all. And we should have been charging for our websites from day one. Subscriptions have been part of this forever.

You have to pay for paper. You have to pay for pixels. It costs money. And I think it was a huge mistake. I can't put that back in the tube now. But if you look at the Internet, the only thing that's free is what we do: information. Everything else costs money. Ring tones cost a dollar. You pay for your access to the Internet. You pay for your e-mail. Everybody says e-mail is free. It's not free. First of all, you're paying your ISP for it. And if you're using something like Google mail, you're turning yourself into an advertising conduit for a giant corporation. There's nothing free about the Internet. It's just baloney.

Putting aside some of the more obvious problems with Rosenthal's logic (if he's a got a problem with being an advertising conduit for giant corporations, he's working for the wrong company), he does touch on some important themes. I quote him in entirety first because I imagine a lot of heads nodded in agreement when they originally read it (and may have nodded again just reading it here). He's also making some assumptions I hear every day: "no price means no value," "you have to pay for pixels," there's nothing free about the Internet because "you're paying your ISP for it," the only thing free on the Internet is "information," and so on.

There is a grain of truth to each of these, but this thinking is also just wrong—deeply, almost head-scratchingly wrong. And yet not a day goes by when I don't hear these and other doubts about Free that are just as ill-conceived (aside from the usual "freetard" label, which I continue to find funny even though I know I shouldn't). So here are the fourteen most frequently heard objections to the notion of an economy based on Free, with an example of each, and my response:

1. There ain't no such thing as a free lunch.

People with a sound education in economics know that nothing is really "free": you're going to pay one way or another for whatever you get.
—Terry Hancock, writing in to complain to
the Free Software Magazine

This chestnut is known as TANSTAFL in the economics world and was a favorite rejoinder of Milton Friedman, the Nobel Prize–winning former University of Chicago economics professor. It simply states that a person or a society cannot get something for nothing. Even if something appears to be free, there is always a cost to the person or to society as a whole, even though that cost may be hidden or distributed.

Is this always true? That's really two questions, an economic one and a practical one: Intellectually, we think that someone has to be paying somewhere, right? And it's going to ultimately be me, isn't it?

The short answer to the first question is yes. Eventually all costs must be paid. What's changing, however, is that those costs are moving from the mostly "hidden" (the small matter of the beer you must buy for that lunch) to the "distributed" (somebody's paying, but it's probably not you; indeed, the costs may be so distributed that we individually don't feel them at all).

Economists usually consider this rule in the context of "closed markets," such as the balance sheet of the restaurant serving that lunch. If you aren't paying for the lunch, your lunch partner is. And if she's not paying, then the restaurant owner is paying. Or if the restaurant owner isn't paying, the food supplier is. And so on. One way or another, the books must balance.

But the world is full of markets that are not closed and tend to leak into the other markets around them, which we may or may not be measuring. We've already looked at the interaction between the monetary and the nonmonetary markets. Lunch may have been free to you in the monetary market, but you paid with your time and presence in the attention and reputation markets. These are "other costs," and are the way economists deal with things that don't fit the basic models. One of those is "opportunity costs"—the value of something else you might have done with the time you were at lunch.

If we could possibly do the accounting over all markets, monetary and nonmonetary, and work out the proper conversion ratios, it would no doubt be true that Friedman was right. But we can't. And even if we could calculate the distributed ecological cost of that breath you just took, for example, it wouldn't take away from the practical reality that it's free to you in every way you care about.

Economics, at least in its idealized form, obeys conservation laws: What goes in must come out. If you print more money, for instance, standard monetary theory says you devalue the existing supply of money by an equivalent amount.

In reality, however, economics is called a "dismal science" for good reason—like other studies of human behavior, it is more than a little fuzzy. What cannot be directly measured in economic systems is hand-waved away into a category called "externalities" (for example, when you buy a pair of shoes you are not charged for the environmental impact of the carbon released in their manufacture—that's called a "negative externality," which we'll discuss at length below). A lot of the costs in that free lunch fall under the category of externalities—technically there, but immaterial to you.

To demonstrate, let's try to follow the money as you pay for reading a Wikipedia entry. The Wikipedia Foundation, which pays for the servers and bandwidth that Wikipedia runs on, is a nonprofit supported by donors, both corporate and individual. Assuming you are not one of those individual donors (and only a minute fraction of Wikipedia's users are), perhaps you are a customer of one of Wikipedia's corporate donors, such as Sun Microsystems. In that case, you may be paying a tiny fraction of a cent more for Sun servers than you would otherwise, to pad Sun's profit margin enough that it can make a charitable donation. Not a Sun customer? Well, Google is a Wikipedia donor, too. Perhaps you once paid for a Google ad that was a zillionth of a cent more expensive than it otherwise would have been had Google not made the donation. Not an advertiser? Well, then maybe you bought a product from one of Google's advertisers, and that product was a gazillionth of a cent more expensive because of this chain of events.

At this point we're talking about fractions of a cent that are like an atom in that penny. In other words, although you can probably argue that you are ultimately paying for that Wikipedia entry, it is only true in the sense that the flutter of a butterfly wing in China could determine your weather tomorrow. Technically, there may be a connection, but it is too small to measure, and so we don't bother.

Let's now revisit Rosenthal's comment that, like paper, "you have to pay for pixels." Well, technically that's true, but as the editor of an operation that both prints paper and pushes pixels, I know the differences are far greater than the similarities. We pay dollars to print, bind, and mail a magazine to you (that's not including any of the cost to produce the content inside), but just microcents to show it to you on our Web site. That's why we can treat it as free, because on a user-by-user basis it is, in fact, too cheap to meter.

Overall, our server and bandwidth bill amounts to several thousand dollars a month. But that's to reach tens of millions of readers. Compared to the value of those readers, we're more than happy to treat pixels as free. Sure, let's grant the naysayers the semantic point: Free isn't really free. But in many cases, it might as well be. That's what matters most in determining how we run our lives and our businesses.

2. Free always has hidden costs/Free is a trick.

Free isn't what it used to be, especially on the Internet, whose very history and technology are based on the notion that information and pretty much everything else online want to be free. Web giveaways increasingly come at a steep price, in the form of computer glitches, frustration and loss of privacy and security—not to mention the threat of expensive lawsuits for large-scale music downloaders.
—*John Schwartz*, New York Times

This is not so much a fallacy as a stereotype. Yes, it's true that Free sometimes comes with strings attached. Advertising clutters your page. Limits are imposed. You're upsold to different products or locked into something very much not free. We are baited, then switched.

But that describes twentieth-century Free a lot better than twenty-first-century Free. Generally, common sense is a good guide: If something seems too good to be true, it probably is, especially in the world of atoms. The marginal costs of a free spritz of perfume in a department store are low enough to believe that it's really free. In contrast, you're right to assume you're going to end up paying for a free vacation, one way or another.

However, in twenty-first-century Free, which is based on the economics of digital bits, there is no need for hidden costs. They might still be there, as Schwartz likes to remind us, but that speaks more to the fact that free products come without warranties, which is the cost of having no guaranteed fix when things go wrong (call that caveat non-emptor).

They are not required by the model. Free can be as good as Paid, or better: no tricks, catches, or strings required (think open source software).

It's time to stop treating bits like atoms and assuming that the same limitations still hold. Trickery is no longer an essential part of the model.

3. The Internet isn't really free because you're paying for access.

Excuse me but when has the internet been free? Free as in speech yes but not free as in beer. We all have to pay an ISP to access the net so we are paying already for what's on it.
—comment on a post by Laurie Langham

This is a common confusion, that somehow the $30 or $40 we spend each month on our Internet access subsidizes the entire Internet. It actually does help pay for the transmission infrastructure, but it has nothing to do with what travels over it. In the same way that minutes of cell phone use say nothing about the value of what's said in those minutes, you're paying for the bits to be delivered to you but not for the value that's in the bits themselves. This is the difference between "content" and "carriage," which are separate markets. Carriage is not free, but content often is. Your monthly ISP (Internet service provider) charge covers the delivery of that content, but the creation of the content is controlled by an entirely different economic model.

It's easy to see why people are confused by this, because there *are* a few markets where carriage subsidizes content, too. In cable TV, for instance, the local cable company pays a license fee for much of the video it sends downs its lines, and that fee comes out of your monthly bill. But the Internet doesn't work that way: Your ISP doesn't control or pay for the bits it transmits. (In legal terms, it's a "common carrier," like a phone company.)

In commonsense terms, this error also comes from measuring the value of a thing using the wrong units. In terms of his mineral content, my youngest son is worth around $5 at current spot market prices, but I won't sell him to you for that. He's worth more to me because of the way those minerals are put together, and all the other atoms, quantum states, and puppy dog tails that combine to make him a person. Confusing the cost of transmitting the megabits with the cost of making them and what they're worth to the receiver is a similar mistake of misunderstanding where the value actually resides. It's not in the network. It's in the production and consumption at the edges, where we turn bits into meaning.

4. Free is just about advertising (and there's a limit to that).

In today's "free" world, in most online business categories, it is inherently impossible to start a small self-sustaining business and to grow it. This is because in the digital world, advertising, the only real revenue stream, cannot support a small digital business. If businesses were based on the idea that people paid for services then small companies could succeed at a small scale and grow. But it is very hard to charge when your competition is free.

—*Hank Williams, writing in* Silicon Alley Observer

One of the biggest fallacies of Free on the Web is that it's only about advertising. Although it's true that advertising-based models dominated the first era of the Web, today freemium—the model where some people pay directly, and support many others who pay nothing at all—is growing fast to rival it (as we saw in Chapter 1). Online video games, for example, are primarily choosing the freemium strategy, as is the fast-growing category of Web-based software (known as "software as a service"). Williams is right that most Web businesses are small ones and it's hard to support a small business on advertising. But he's wrong that this is the only business model available to companies online. More and more companies are like Chicago's 37signals, which uses free samples and trials to market software that is paid for the old-fashioned way: direct payment by customers.

Sound very old-economy? Perhaps, but with the right market it can work. David Heinemeier Hansson, one of 37signals' founders, says the company's secret is not to target individual consumers (it's hard to get them to open their wallets) or big companies (that's a crowded space and a slow purchasing process). Instead, the company focuses on the "Fortune 5 Million"—small companies with specific needs that are underserved. Project management software for twelve-person teams is the kind of product 37signals sells, and the companies who are frustrated by overbuilt one-size-fits-all software from big companies and hard-to-use open source alternatives are happy to pay a few hundred dollars a year for the 37signals version.

You only have to look to the iPhone's "app market" (all the programs you can download from iTunes for your phone) to see hundreds of other small businesses (many of them single programmers) making a tidy income by selling software in a market where others give the software away. No advertising required—it's a straight sale, either up front or for the advanced version of a free basic form. The same is true for thousands of companies that distribute narrowly targeted utility programs online, which tend to be free to try. Nothing new about this—the shareware market has been around for decades—but as this software becomes Web-based it gets easier and easier to do.

The second objection to the assertion that Free online is all about advertising is that this implies Free must be limited: Surely the advertising pie can only be so big. Although that may be true, it's not at all clear where the limits lie and how much bigger the advertising market can get online. Google has shown that online ads can be different enough—measurable, targeted, paid only for performance—to attract an entirely new class of advertiser: small and medium-sized businesses buying keywords for pennies per click. Google isn't just taking more of the advertising pie—it's also making a bigger pie.

5. Free means more ads, and that means less privacy.

*I have asked many people I know who live and die by Facebook how much they would pay for it, and they all said **zero**. So Facebook becomes a slave to advertising and pimping out their users' information for every cent they can get. It isn't unrealistic to think that if people paid for more services that their personal information wouldn't be shared quite so freely.*

—*Paul Ellis, pseudosavant.com*

This is a commonly heard concern about advertising in general. People often assume that any site with advertising must be tracking user behavior and selling that information to the advertisers. This argument fuels the assumption that because so much Free is ad-supported, Free is a corrosive force, spreading marketing sneakiness everywhere it goes.

In truth, the hypothetical Facebook example is still more the exception than the rule. Most ad-driven sites have privacy policies that forbid passing user information to advertisers (and most advertisers wouldn't know what to do with that information even if they got it).

Ellis argues that paying for services directly, rather than having advertisers pay for you, means that the sites will be more inclined to protect your privacy—i.e., they work for you, not the advertisers. While that may be true, it's not necessary.

The media world has been figuring out how to balance consumer and advertiser interests for decades, ranging from industry guidelines to the "Chinese walls" of media that separate editorial and business functions. It is not a new problem, and we've seen that it's possible to have editorial independence even when the advertisers are picking up the freight.

It's worth noting, however, that privacy is a moving target. In Europe, a vast system of laws protects personal information, but in the United States it's more a matter of individual company codes of conduct and consumer pressure. But the expectations of privacy that we had twenty years ago are not mirrored by the generation growing up online today. After you've "overshared" pictures of the drunken scene at your last frat party and described the ups and downs of your latest love affair, how much worse is it if a marketer sends you a discount on a clothing line based on your listed preferences?

6. No cost = no value

> I'm sad that people feel like music should be free, that the work that we do is not valued. When music comes free by way of friends burning CDs, there's not that understanding of the work that goes into the making of an album.
>
> —Sheryl Crow, interviewed in the New York Times Magazine

Spot the fallacy? It's that the only way to measure value is with money. The Web is built mostly on two nonmonetary units—attention (traffic) and reputation (links)—both of which benefit hugely from free content and services. And it's a pretty simple matter to convert either of those two currencies into cash, as a glance at Google's balance sheet makes clear. Or take the TED conference, which charges thousands of dollars per ticket while simultaneously broadcasting all of its talks online (see the sidebar on page 117). Crow, of course, benefits in reputation from her fans ripping her albums and burning them for more friends. The gift of a CD is a recommendation from a trusted party. There are marketers who would kill to facilitate such authentic and rapid word of mouth.

When people rip and burn CDs (or, more probably these days, share music electronically via iTunes or file trading), they're not saying that Crow didn't put any work into the album. They're essentially saying she didn't put any work into that particular act of distribution—the creation of a digital copy. And, indeed, she didn't. The marginal cost to her of that transfer is zero, and the file-trading generation's innate understanding of digital economics helps usher in the conclusion that her payment for that transfer should also be zero.

Crow will make her money in the end, through her concerts, her merchandise, licensing for commercials or soundtracks, and yes, the sale of some of her music to people who still want CDs or prefer to buy their music online. But the celebrity and credibility that she gets from the file traders who choose to download her music or the CD swappers who opt to rip and burn will help—at the very least, file sharing wins her reputational currency. It's impossible to quantify how much of that will translate to cash through these other means, but it's not zero. Is it more than the amount of direct revenues she might get if these people paid for the music? We'll never know.

At the end of the book, I list fifty business models built on Free, and there are hundreds more. All of them are based on the notion that free stuff does have value and the way we measure that is through people's actions. There is no greater test of what people value than what they choose to spend their time on—although we are getting more affluent, we're not getting any more hours in the day. Crow is being listened to by the most distracted generation in history, with the most choice and the most competition for their time. There are worse problems than getting attention.

7. Free undermines innovation.

Of the world's economies, there's more that believe in intellectual property today than ever. There are fewer communists in the world today than there were. But there are some new modern-day sort of communists who want to get rid of the incentive for musicians and moviemakers and software makers under various guises.
—Bill Gates, *interviewed in 2005*

The argument that Free attacks intellectual property rights such as patents and copyright straddles the line between *libre* and *gratis*. The thinking goes like this: People aren't going to invent things if they can't be rewarded for it. Patents and copyright are our way of ensuring that creators get paid. So what's the point of patents and copyright if the marketplace expects the price to be zero?

Actually, the history of intellectual property law fully recognizes the power of Free. It's based on the long traditions of the scientific world, where researchers freely build on the published work of those who came before. In the same vein, the creators of the patent system (led by Thomas Jefferson) wanted to encourage sharing of information, but they realized that the only way people thought they could get paid for their inventions was to hold them secret. So the Founding Fathers found another way to protect inventors—the seventeen-year patent period. In exchange for open publication of an invention (*libre*), the inventor can charge a license fee (not *gratis*) to anyone who uses it for the term of the patent. But after that term expires, the intellectual property will be free (*gratis*).

So there's already a place for Free in patents—it kicks in after seventeen years. (Copyright was also meant to expire, but Congress keeps extending it.) However, a growing community of creators doesn't want to wait that long. They're choosing to reject these rights and release their ideas (whether as words, pictures, music, or code) under licenses such as Creative Commons or various open source software licenses. They believe that real Free—both *gratis* and *libre*—encourages innovation by making it easier for other people to remix, mash up, and otherwise build on the work of others.

As for making money, they do so indirectly, either in selling services around the free goods (such as supporting Linux) or in just finding ways to turn the reputational currency they've earned by having others build on their work (with due credit) into cash through better jobs, paid gigs, and the like.

8. Depleted oceans, filthy public toilets, and global warming are the real cost of Free.

Free parking has contributed to auto dependence, rapid urban sprawl, extravagant energy use, and a host of other problems. Planners mandate free parking to alleviate congestion, but end up distorting transportation choices, debasing urban design, damaging the economy, and degrading the environment. Ubiquitous free parking helps explain why our cities sprawl on a scale fit more for cars than for people, and why American motor vehicles now consume one-eighth of the world's total oil production.
    —*Donald Shoup*, The High Cost of Free Parking

No discussion of Free can avoid "The Tragedy of the Commons." If we don't have to pay for things, we tend to consume them to excess. The classic tragedy of the commons example (which biologist Garrett Hardin used in a 1968 article) is sheep grazing on the commonly owned village green. Since sheep owners don't have to pay for the land, they are not incentivized to preserve it. Indeed, it is even worse: Since they know that others are similarly able to waste the resource, they may choose to gain a bigger share of the benefit by wasting it *faster,* grazing more of their sheep, more of the time, until quickly the green is brown.

This is the consequence of what economists call "uncompensated negative externalities." When things are actually scarce (limited) but we price them as if they were abundant (essentially unlimited), bad things can happen.

Take global warming. We now realize that the cost of putting tons of carbon into the atmosphere is that global temperatures will rise and create all sorts of dire consequences. But we had priced atmospheric carbon release as if there were no consequences, which is to say we didn't price it at all. You were free to release as much carbon as you wanted into the atmosphere, and as a result we released as much as we could. In other words, the environmental cost of carbon was both "external" to our economic system and, as it turned out, negative. Current efforts to impose carbon taxes, caps, and other limits are attempts to compensate for those costs by making them "internal" to our economic system.

You can see this problem all around you. We've overfished the oceans because limits either don't exist or are unenforced and fishermen treat fish as "free." On the personal level, if you walk into a disgusting public toilet, you can probably smell uncompensated negative externalities. The toilet is free to use and the cost of cleaning it up is borne by someone else, so people tend not to treat it with the same care they would treat their own toilets, where the costs are felt directly. And so on, from litter to deforestation. Free can lead to gorging, and thus ruin the party for everyone.

But note that the environmental costs of Free are mostly felt in the world of atoms. As we've discussed, it's hard to make atoms really free— the reason that we don't feel these environmental costs is simply that we haven't priced the market right. Those plastic bags are only free because we don't charge *directly* for the cleanup costs of taking them out of trees. But increasingly, we are starting to measure and account for the negative externalities (making them negative *internalities,* since they're now explicitly part of a closed economic system). As such, you're starting to see either supermarket discounts for using nondisposable bags (which is effectively the same as a surcharge for plastic bags) or bans on plastic bags altogether.

In the world of bits, the environmental costs are far less of an issue. Wasteful use of processing, storage, and bandwidth basically boils down to electricity, and the market is getting increasingly good at pricing the environmental costs of that. Carbon caps, regulation requiring renewable energy sources, and local emission limits have pushed companies such as Google, Microsoft, and Yahoo! to locate their data centers near hydroelectric power sources, which are carbon-free. Eventually, they will be located near solar-thermal, wind, and geothermal electricity sources, too. Simple economics—regulation making carbon-generating electricity more expensive than renewable electricity—will ensure that wasting bits will not have the environmental consequences of wasting atoms.

But digital Free can have unaccounted costs, too. Consider flat-fee broadband access, where incremental use is free. (This is the standard plan for your cable modem or DSL connection.) Some people change their behavior with such free capacity and share huge files with peer-to-peer file-trading software such as BitTorrent. This minority ends up using the majority of the network capacity, and as a result all of our Internet access is slower.

That's why Internet service providers such as your cable company cap individual users who use too much capacity, or charge more for people who want to transfer more data. It's typically a pretty high cap that doesn't affect most of us, and the ISPs are careful to keep it that way. But given that most of us have a choice of broadband providers, few ISPs want to get a reputation as the "slow" one.

9. Free encourages piracy.

You're a communist aren't you Mike? I can tell you are sure not a capitalist. Content creators should be paid for content they make. I will admit right here and now I download content and I will continue to thieve content anywhere I can get it. But I don't try to rationalize my immorality with rhetoric about "economics when there's a lack of scarcity."
　　　　—*"Xenohacker," responding to Mike Masnick at Techdirt*
　　　　[spelling vastly improved]

No, it's just the opposite. Free doesn't encourage piracy. Piracy encourages Free. Piracy happens when the marketplace realizes that the marginal cost of reproduction and distribution of a product is significantly lower than the price asked. In other words, the only thing propping up the price is the law protecting intellectual property. If you break the law, the price can fall, sometimes all the way to zero. That's true for everything from fake Louis Vuitton luggage (where the price is low, but not zero) to MP3s (which are traded without charge).

So piracy is like the force of gravity. If you're holding something off the ground, sooner or later gravity is going to win and it will fall. For digital products it's the same thing—copyright protection schemes, coded into either law or software, are simply holding up a price against the force of gravity. Sooner or later, it will fall, either because the owner drops it or because the pirates knock it to the ground.

This is not to condone or encourage piracy, only to say that it is more like a natural force than a social behavior that can be trained or legislated away. The economic incentives to pirate digital goods—zero cost, identical reward—are so great that it can be assumed that anything of value in digital form will eventually be pirated and then freely distributed. Sometimes that stays within a shadowy subculture (enterprise software piracy), and sometimes it goes mainstream (music and movies). But it is almost impossible to stop. Economics has little place for morality for the same reason that evolution is unsentimental about extinction—it describes what happens, not what *should* happen.

10. Free is breeding a generation that doesn't value anything.

Just a few decades ago, people had low expectations and worked hard to make a living. They did not know free and never expected it. Now, the opposite trend is happening, with free becoming expected online. Will the new generation, the one that expects something for nothing, work as hard to maintain the high standards of living that we created?
—Alex Iskold, ReadWriteWeb

This has been a worry since the Industrial Revolution—replace "free" with "steam" and you can imagine the Victorian concern about flabby muscles and minds. It is true that each generation takes for granted some things their parents valued, but that doesn't mean that generation values *everything* less. Instead, they value *different things*. Somehow we managed to stop getting up at dawn to milk the cows without losing our overall will to work.

It's true that anyone growing up in a broadband household today will likely assume that everything digital should be free (probably because almost all of it is). Call them Generation Free.

This group—most people under the age of twenty in the developed world—also expects information to be infinite and immediate. (They're otherwise known as the Google Generation.) They are increasingly unwilling to pay for content and other entertainment, because they have so many free alternatives. This is the generation that wouldn't think of shoplifting but doesn't think twice about downloading music from file-trading sites. They intuitively understand the economics of atoms versus bits, and realized that the first has real costs that must be paid, but the second usually does not. From that perspective, shoplifting is theft but file trading is a victimless crime.

They insist on Free not just in price but also in the absence of restrictions: They resist registration barriers, copyright control schemes, and content that they can't own. The question is not "What does it cost?" but "Why should I pay?" This is not arrogance or entitlement—it is experience. They have come of age in a world of Free.

When I explain the thesis of this book—that making money around Free will be the future of business—the response from this generation is usually "And?" It seems self-evident to them. This is the difference between digital natives and the rest of us. They somehow understand near-zero marginal costs from birth (although not perhaps in those words).

But Generation Free doesn't assume that as go bits, so should go atoms. They don't expect to get their clothes or apartments for free; indeed they're paying more than ever for those. Give the kids credit: They can differentiate between the physical and the virtual, and they tailor their behavior differently in each domain. Free online is no more likely to lead to an expectation of free offline than to lead to an expectation that people should look like their World of Warcraft characters.

11. You can't compete with free.

There is no business model ever struck off by the hand and grain of man that can compete with free. It can't be done. If I have a Pizza Hut and I'm selling pizzas at $1.50, somebody puts up a Pizza Hut next to me and gives them away, who do you think is going to get the business?
    —Jack Valenti, Motion Picture Association of America

Chapter 14 is all about this, looking particularly at how Microsoft learned to compete with open source software. But the short form is that it's easy to compete with Free: simply offer something better or at least different from the free version. There is a reason why office workers walk past the free coffee in the kitchen to go out and spend $4 for a venti latte at Starbucks—the Starbucks coffee tastes better. There's also a bit of consumer psychology going on—the small treat, the ritual of pampering ourselves with something a little luxurious. It's easy to find free coffee, but what Starbucks offers is something more.

The situation is rarely the one Valenti postulates: a free Pizza Hut opening next to a regular one. Instead, it's far more likely to be something like a Domino's, which offers free pizza if the delivery takes longer than thirty minutes, opening up next to a Pizza Hut. These are different services, and Free is just one of many factors in deciding between them.

The way to compete with Free is to move past the abundance to find the adjacent scarcity. If software is free, sell support. If phone calls are free, sell distant labor and talent that can be reached by those free calls (the Indian outsourcing model in a nutshell). If your skills are being turned into a commodity that can be done by software (hello, travel agents, stockbrokers, and realtors), then move upstream to more complicated problems that still require the human touch. Not only can you compete with Free in that instance, but the people who need these custom solutions are often the ones most willing to pay highly for them.

12. I gave away my stuff and didn't make much money!

What is the writer or musician to do, though, if she can't earn money from her art? Simple, says the Slashdotter: earn your money playing live (if you're one of those musicians who plays live), or selling T-shirts or merchandise, or providing some other kind of "value-added" service. Many such arguments seem to me to be simple greed disguised in highfalutin' idealism about how "information wants to be free."
—Steven Poole, *author of* Trigger Happy

Steven Poole wrote a terrific book about video game culture a half decade ago, but unfortunately he more recently ran a rather halfhearted experiment in Free. In 2007, after his book was no longer being sold in most bookstores, he posted it as a file on his blog. He also added a tip jar so people could give him money if they wanted. Few did—just one out of every 1,750. So he declared that giving away free books was a bust.

It is indeed true that his particular experiment was a failure, but that says more about the exercise than it does about Free. Putting a tip jar next to free goods is what Mike Masnick of Techdirt, a news and analysis site, calls the mistake of "give it away and pray." Rather than being a failed Free business model, it's no business model at all.

What's a better model? Well, for starters, giving away a book closer to the time of publication, not only years afterward. Take Paulo Coelho. Total sales of his books topped more than 100 million in 2007, something he credited in part to the buzz he got by putting his most popular book, *The Alchemist,* and dozens of translations of his newer books, on free peer-to-peer file-trading services such as BitTorrent.

Initially, his publisher, HarperCollins, had been against the idea of the author self-pirating his book. So Coelho set up a fake blog, Pirate Coelho, ostensibly written by a fan "liberating" the works. It got attention, and even his older books returned to the *New York Times* bestseller list. When his next book, *The Witch of Portobello,* was published in 2007, he did it again, and it, too, became a best seller.

That, in turn, got HarperCollins's attention. It decided to give away a new Coelho book each month on its own site (albeit only for a month each, and in a somewhat crippled form that didn't allow printing).

"I do think that when a reader has the possibility to read some chapters, he or she can always decide to buy the book later," Coelho said in an interview. "The ultimate goal of a writer is to be read. Money comes later."

Even less-well-known authors can use Free effectively. Matt Mason, who wrote *The Pirate's Dilemma,* used a name-your-own-price model (with zero being an option) for ebooks. When you go through the checkout process, the default price is $5 via PayPal. Of the nearly 8,000 people who downloaded the book, about 6 percent paid, with an average price of $4.20. That adds up to only a couple thousand dollars of direct revenues, but he estimates that the attention he got from the exercise earned him another $50,000 in lecture fees.

Compared to these examples, a digital tip jar next to the free PDF of a seven-year-old book is a joke (sorry, Poole!). From where I sit, it seems bizarre that Poole wasn't delighted to have found 32,000 new readers for a book at the end of its life. If, at that point, he can't then turn that increased readership into some kind of indirect revenue stream, whether speaking, teaching, more writing, consulting, or just more traffic to his blog, he's not as smart as I think he is.

Free is not a magic bullet. Giving away what you do will not make you rich by itself. You have to think creatively about how to convert the reputation and attention you can get from Free into cash. Every person and every project will require a different answer to that challenge, and sometimes it won't work at all. This is just like everything else in life— the only mystery is why people blame Free for their own poverty of imagination and intolerance for possible failure.

13. Free is only good if someone else is paying for it.

We don't want to waste our time with a product or service if it's not worth anything. We want things of value, and we don't want to waste a lot of time trying to determine if what is being offered is something we would use or consume. The easiest way to make the determination? See if anyone else is using it and paying for it.
                    —Mark Cuban, billionaire technology entrepreneur and owner
                                                    of the Dallas Mavericks

Cuban has a point: We often do apply a relative sense of value to free things depending on what we think their market price is. Just as a "50 percent off" sale can entice you to buy something you don't really want because you can't resist the savings, getting something free when others are paying can make that thing look more attractive—gratis-tinted glasses.

But this is more the exception than the rule, for two reasons. First, the ascendant freemium model satisfies Cuban's challenge perfectly without actually conforming to his specific construction. In the case of freemium, someone else is paying, but they're paying for a premium version of the product you're getting for free. That shows that even if the free version hasn't passed the wallet test, its cousin has and you can feel confident in the lineage. For instance, you may feel that Google Earth is of professional quality because it's related to the very expensive Google Earth Pro, just with some features removed. (In fairness, Cuban acknowledges that freemium satisfies his criteria, but because he overstated his thesis and so many people seem to agree with him, I'll continue to poke holes.)

The second reason Cuban's point only goes so far is there are so many counterexamples. Nobody thinks less of Facebook because it's free or longs for a Web browser that people are paying for. When something used to cost money and is now free, you might think less of it—a formerly hot club now letting in anyone gratis. But if something has always been free and there is no expectation otherwise, there's little evidence that people view it with less regard. Web sites are evaluated on their merits, and people have learned that a pay site is actually more likely to be a rip-off than a free one, since it can steal more than just time.

14. Free drives out professionals in favor of amateurs, at a cost to quality.

> *It is no coincidence that just as you have the rise of The Huffington Post that encourages people to give away their content for free you have job losses and the death of the professional journalist.*
> —Andrew Keen, author of The Cult of the Amateur

It's true: Free does tend to level the playing field between professionals and amateurs. As more people create content for nonmonetary reasons, the competition to those doing it for money grows. (As the employer of lots of professional journalists, I think about the relative roles of the amateurs and the pros all the time.) All this means is that publishing is no longer the sole privilege of the paid. It doesn't mean that you can't get paid for publishing.

Instead, the professional journalists who are seeing their jobs evaporate are typically those whose employers failed to find a new role in a world of abundant information. By and large, that means newspapers, which are an industry that will probably have to reinvent itself as dramatically as music labels. The top tier (the *New York Times, Wall Street Journal,* etc.) will probably shrink a bit, and the tier below that may be decimated.

But out of the bloodbath will come a new role for professional journalists. There may be more of them, not fewer, as the ability to participate in journalism extends beyond the credentialed halls of traditional media. But they may be paid far less, and for many it won't be a full-time job at all. Journalism as a profession will share the stage with journalism as an avocation. Meanwhile, others may use their skills to teach and organize amateurs to do a better job covering their own communities, becoming more editor/coach than writer. If so, leveraging the Free— paying people to get *other* people to write for nonmonetary rewards— may not be the enemy of professional journalists. Instead, it may be their salvation.

CODA

Free in a Time of Economic Crisis

AFTER THE 2001 STOCK MARKET CRASH, the business models of the dot-com economy were laid bare. How silly we had been to believe that "monetizing eyeballs" was a good foundation for a business! What were we investors thinking when we stuffed our portfolios with shares of companies that sold pet food online? "Amazon.Bomb," sneered the headlines. We hung our heads in shame at our embrace of some fanciful "new economy."

A few years later, when the markets recovered and we looked back, we found to our surprise that it was practically impossible to see the effect of the crash on the growth of the Internet. It had continued to spread, just as before, with hardly a dip as the public markets cratered. The "digital revolution" hadn't been a mirage, or worse, a hoax. The number of people getting online had climbed at the same rate throughout, as had traffic, and pretty much every other measure of impact.

This had been a Wall Street bubble, not a technology one. The Web was every bit as important as even the most starry-eyed forecasters had predicted—it just took a bit longer to get there than the stock market multiples had assumed.

Now the markets have crashed again. Will Free be more like Web traffic and grow regardless, or more like online pet food?

From a consumer perspective, Free is far more attractive in a down economy. After all, when you have no money, $0.00 is a very good price. Expect the shift toward open source software (which is free) and

Web-based productivity tools such as Google Docs (also free) to accelerate. The cheapest and coolest computers today are "netbooks," which
sell for as little as $250 and ship with either free versions of Linux or
supercheap old versions of Windows. The people who buy them don't
load Office and pay Microsoft hundreds of dollars for the privilege. Instead, they use online equivalents, as the netbook name implies, and
those tend to be free.

These same consumers are saving their money and playing free online games, listening to free music on Pandora, canceling basic cable
and watching free video on Hulu, and killing their landlines in favor of
Skype. It's a consumer's paradise: The Web has become the biggest
store in history and everything is 100 percent off.

What about those companies trying to build a business on the Web?
In the old days (that would be until September of 2008) the model was
pretty simple: (1) Have a great idea; (2) Raise money to bring it to market, ideally free to reach the largest possible audience; (3) If it proves
popular, raise more money to scale it up; (4) Repeat until you're bought
by a bigger company.

Now steps 2 through 4 are no longer available. So Web start-ups are
having to do the unthinkable: come up with a business model that
brings in real money while they're still young.

This is, of course, nothing new in the world of business. But it is a bit
of a shock in the Web world, where attention and reputation are the currencies most in demand, with the expectation that a sufficient amount of
either will turn into money someday, somehow.

The standard business model for Web companies that don't actually
have a business model is advertising. A popular service will have lots of
users, and a few ads on the side will pay the bills. Two problems have
emerged with that model: the price of online ads and click-through rates.
Facebook is an amazingly popular service, but it is also an amazingly in-
effective advertising platform. Even if you could figure out what the
right ad to serve next to a high school girl's party pictures might be, she
and her friends probably won't click on it. No wonder Facebook applications get less than $1 per 1,000 views (compared to around $20 on big
media Web sites).

Google has built an enviable economic engine on the back of its targeted text ads, but the sites on which they run rarely feel as flush. Running Google's AdSense ads on the side of your blog, no matter how popular it may be, will not pay you even minimum wage for the time you spend writing it. On a good month it might cover your hosting fees. I speak from experience.

What about the oldest trick in the book: actually charging people for your goods and services? This is where the real innovation will flourish in a down economy. It's now time for entrepreneurs to innovate, not just with new products but with new business models.

Take Tapulous, the creator of Tap Tap Revenge, a popular music game program for the iPhone. As in Guitar Hero or Rock Band, notes stream down the screen and you have to hit them on the beat. Millions of people have tried the free version, and a sizable fraction of them were ready and willing to pay when Tapulous offered paid versions built around specific bands, such as Weezer and Nine Inch Nails, along with add-on songs. (The *Wall Street Journal* is pursuing a strategy of blending free and paid content on its Web site.)

At the other end of the business spectrum there's Microsoft, which now has to compete with the free word processors and spreadsheets of online competitors such as Google. Rather than complain about the unfair competition (which would be ironic), Microsoft created Web versions of its business software and offered them free to small and young companies. If your firm is less than three years old and under $1 million in revenues, you can use Microsoft's software without charge under its BizSpark program. When those companies get bigger, Microsoft is betting that they'll keep using its software as paying customers. In the meantime, the program costs Microsoft almost nothing.

But extracting a business model from Free is not always easy, especially when your users have come to expect gratis. Take Twitter, the fantastically popular (and free, of course) 140-character messaging service where people update the world on what they're doing, one haiku-like snippet at a time. After taking over the world, or at least the geeky side of it, it now finds itself having to actually make enough money to cover its bandwidth bills. Last year it hired a revenue guru to try to find a business model and has announced that it intends to reveal its strategy in 2009. Speculation as to what that will be ranges from charging companies to have their "tweets" recommended to consumers (which is a bit like "friending" the Burger King on Facebook) to certifying identity to avoid impersonation. The revenue officer has his work cut out for him.

Meanwhile YouTube is still struggling to match its popularity with revenues, and Facebook is selling commodity ads for pennies after its effort to charge for intrusive advertising led to a user backlash. And news-sharing site Digg, for all its millions of users, still doesn't make a dime. A year ago, that hardly mattered: The business model was "build to a lucrative exit, preferably in cash." But now the exit doors are closed and cash flow is king.

Does this mean that Free will retreat in a down economy? Probably not. The psychological and economic case for it remains as good as ever—the marginal cost of anything digital falls by 50 percent every year, making pricing a race to the bottom, and "Free" has as much power over the consumer psyche as ever. But it does mean that Free is not enough. It also has to be matched with Paid. Just as King Gillette's free razors only made business sense paired with expensive blades, so will today's Web entrepreneurs have to invent not just products that people love but also those that they will pay for. Free may be the best price, but it can't be the only one.

FREE RULES

The Ten Principles of Abundance Thinking

1. **If it's digital, sooner or later it's going to be free.**
    In a competitive market, price falls to the marginal cost. The Internet is the most competitive market the world has ever seen, and the marginal costs of the technologies on which it runs—processing, bandwidth, and storage—get closer and closer to zero every year. Free becomes not just an option but an inevitability. Bits want to be free.

2. **Atoms would like to be free, too, but they're not so pushy about it.**
    Outside of the digital realm, marginal costs rarely fall to zero. But Free is so psychologically attractive that marketers will always find ways to invoke it by redefining their business to make some things free while selling others. It's not really free—it's probably you paying sooner or later—but it's often compelling all the same. Today, by creatively expanding the definition of their industry, companies from airlines to cars have found ways to make their core product free by selling something else.

3. **You can't stop Free.**
    In the digital realm you can try to keep Free at bay with laws and locks, but eventually the force of economic gravity will win. What that means is that if the only thing stopping your product

from being free is a secret code or a scary warning, you can be sure that there's someone out there who will defeat it. Take Free back from the pirates, and sell upgrades.

4. **You *can* make money from Free.**
People will pay to save time. People will pay to lower risk. People will pay for things they love. People will pay for status. People will pay if you make them (once they're hooked). There are countless ways to make money around Free (I list fifty of them at the end of the book). Free opens doors, reaching new consumers. It doesn't mean you can't charge some of them.

5. **Redefine your market.**
Ryanair's competitors were in the airline seat business. It decided to be in the *travel* business instead. The difference: There are dozens of ways to make money in travel, from car rentals to subsidies from destinations hungry for tourists. The airline made its seats cheap, even free, to make more money *around* them.

6. **Round down.**
If the cost of something is heading to zero, Free is just a matter of when, not if. Why not get there first, before someone else does? The first to Free gets attention, and there are always ways to turn that into money. What can you make free today?

7. **Sooner or later you will compete with Free.**
Whether through cross-subsidies or software, somebody in your business is going to find a way to give away what you charge for. It may not be exactly the same thing, but the price discount of 100 percent may matter more. Your choice: Match that price and sell something else, or ensure that the differences in quality overcome the differences in price.

8. **Embrace waste.**

   If something is becoming too cheap to meter, stop metering it. From having flat fees to no fees, the most innovative companies are those who see which way the pricing trends are going and get ahead of them. "Your voice mail inbox is full" is the death rattle of an industry stuck with a scarcity model in a world of capacity abundance.

9. **Free makes other things more valuable.**

   Every abundance creates a new scarcity. A hundred years ago entertainment was scarce and time plentiful; now it's the reverse. When one product or service becomes free, value migrates to the next higher layer. Go there.

10. **Manage for abundance, not scarcity.**

    Where resources are scarce, they are also expensive—you have to be careful how you use them. Thus traditional top-down management, which is all about control to avoid expensive mistakes. But when resources are cheap, you don't have to manage the same way. As business functions become digital, they can also become more independent without risk of sinking the mothership. Company culture can shift from "Don't screw up" to "Fail fast."

FREEMIUM TACTICS

FINDING A FREEMIUM MODEL THAT WORKS FOR YOU

There are countless variations of the freemium model, but as an example of how to pick one, consider a business software company that offers its product as an online service. Initially it was charging all of its users from $99 to tens of thousands of dollars a year for the software. But it wanted to use Free to reach more people. Here are four models it considered:

1. **Time limited** (Thirty days free, then pay. This is the Salesforce model.)

    - Upside: Easy to do, low risk of cannibalization.
    - Downside: Many potential customers will be unwilling to commit enough to give the software a real test, since they know that if they don't pay they'll get no benefit after thirty days.

2. **Feature limited** (Basic version free, more sophisticated version paid. This is the WordPress model.)

    - Upside: Best way to maximize reach. When customers convert to Paid, they're doing it for the right reason (they understand the value of what they're paying for) and are likely to be more loyal and less price sensitive.
    - Downside: Need to create two versions of the product. If you put too many features in the free version, not enough people

will convert. If you put too few, not enough will use it long enough to convert.

3. **Seat limited** (Can be used by up to some number of people for free, but more than that is paid. This is the Intuit QuickBooks model.)

   - Upside: Easy to implement. Easy to understand.
   - Downside: Might cannibalize the low end of the market.

4. **Customer type limited** (Small and young companies get it free; bigger and older companies pay. This is the model used by Microsoft's BizSpark, where companies less than three years old and under $1 million in revenues get Microsoft's business software free.)

   - Upside: Charges companies according to their ability to pay. Gets fast-growing companies early.
   - Downside: Complicated and hard-to-police verification process.

In the end, the software company went with time limited, since it was the easiest to implement. But the CEO is still considering the others. The problem with free trials is that they discourage full participation during the trial period: Why spend a lot of time learning to use something when there's a chance that when the time comes to pay, you may not feel it's worth it? Indeed, why start using it at all?

Time-limited freemium models may have relatively high conversion rates to Paid from those who continue to use the product throughout the trial period, but they may be limiting the total number of participants. Some effort in creating a version that offers a more useful experience for the free user, without the risk of being cut off when the clock runs out, can increase the overall reach of the product. Even if a smaller percentage converts, it may be a smaller percentage of a much larger number.

WHAT'S THE RIGHT
CONVERSION PERCENTAGE?

In Chapter 2, I described freemium as the opposite of the traditional free sample: Instead of giving away 5 percent of your product to sell 95 percent, you give away 95 percent of your product to sell 5 percent. The reason this makes sense is that for digital products, where the marginal cost is close to zero, the 95 percent costs you little and allows you to reach a huge market. So the 5 percent you convert is 5 percent of a big number.

But that was just a hypothetical percentage split, to make a point. In the real world, what's the right balance? The answer varies from market to market, but some of the best data is in the games world.

In online free-to-play games, companies aim to structure their costs so they can break even if as little as 5 to 10 percent of the users pay. Anything above that is profit. Which is why these numbers from blogger Nabeel Hyatt, who follows the industry, are so impressive:

- **Club Penguin**: 25 percent of monthly players pay, $5/month per paying user
- **Habbo**: 10 percent of monthly players pay, $10.30/month per paying user
- **RuneScape**: 16.6 percent of monthly players pay, $5/month per paying user
- **Puzzle Pirates**: 22 percent of monthly players pay, $7.95/ month per paying user

As the blog notes, that compares very well to the 2 percent of the casual downloadable game market that pays, or the 3 to 5 percent that a lot of "penny gap" free trial Web start-ups get. Estimates for the number of free Flickr users that convert to paid Flickr Pro range from 5 to 10 percent. And shareware software programs often see less than 0.5 percent of users paying up.

But other companies are able to do much better. Intuit, for instance, offers basic TurboTax Online free for federal taxes, but charges you for the state version. Company officials tell me 70 percent of users opt to pay for that version. That's a special case—practically everyone has to pay both federal and state taxes—but it's evidence that some very high conversion rates are possible in the freemium model.

For the typical Web 2.0 company planning to use freemium as its revenue model, my advice would be to set 5 percent as break-even, but balance the mix of free versus paid features with the hopes of actually converting 10 percent. More than that, and you may be offering too little in your free version and thus not maximizing the reach that's possible with Free. Less than that, and the costs of the freeloaders start to get significant, making it difficult to make money.

WHAT'S A FREE CUSTOMER WORTH?

It turns out that not all free customers are alike, and what they're worth to you depends on when they arrive. In the early stages of a company or product, when it's trying to get traction, Free is the best marketing.

It lowers the risk for new customers to try the product, and it increases the product's potential reach. But over time, as the product or company becomes more established and better known, there's less risk in trying it and Free becomes less essential.

This was quantified by Sunil Gupta and Carl Mela, two Harvard Business School professors who analyzed* an online auction company they called auctions.com (presumably, it was actually eBay). Sellers paid but buyers could use the service for free. The question was what these free buyers were worth.

The answer is: more at the company's start than after it was a few years old. Specifically, the lifetime value of the free buyer who started

---

*Gupta, Sunil, and Carl F. Mela. "What Is a Free Customer Worth?" *Harvard Business Review* 86, no. 11 (November 2008).

using the auction service in its first year was $2,500. As those early adopters, enticed by the free service, brought in other buyers, the critical mass of buyers brought a critical mass of sellers.

Eight years later, with the auction company well established, the lifetime value of a new customer was much less: just $213. They might spend just as much as the early adopters did at the start, but their value wasn't multiplied by a torrent of other users who came with them. The auction company kept the price of participation for buyers at zero, since its costs were pretty close to zero, too. But another company with higher costs might have switched to a pay model once momentum had been achieved. Knowing how the value of a customer changes over time can help you figure out what the right time for Free is, and when it's no longer necessary.

FIFTY BUSINESS MODELS
BUILT ON FREE

THERE ARE COUNTLESS EXAMPLES of Free business models already in action today. Here are fifty examples, organized by the type of Free model they most closely fit into.

FREE 1: DIRECT CROSS-SUBSIDIES

- Give away services, sell products
    (Apple Store Genius Bar tech support)
- Give away products, sell services
    (free gifts when you open a bank account)
- Give away software, sell hardware
    (IBM and HP's Linux offerings)
- Give away hardware, sell software
    (the video game console model, where devices
    such as the Xbox 360 are sold far under cost)
- Give away cell phones, sell minutes of talk time
    (many carriers)
- Give away talk time, sell cell phones
    (many of the same carriers, with free nights and
    weekend plans)
- Give away the show, sell the drinks (strip clubs)
- Give away the drinks, sell the show (casinos)

- Free with purchase (retailer "loss leaders")
- Buy one, get one free (supermarkets)
- Free gift inside (cereal)
- Free shipping for orders over $25 (Amazon)
- Free samples (everything from gift boxes for new mothers to supermarket tasters)
- Free trials (magazine subscriptions)
- Free parking (malls)
- Free condiments (restaurants)

FREE 2: THREE-PARTY, OR "TWO-SIDED," MARKETS (ONE CUSTOMER CLASS SUBSIDIZES ANOTHER)

- Give away content, sell access to the audience (ad-supported media)
- Give away credit cards without a fee, charge merchants a transaction fee
- Give away scientific articles, charge authors to publish them (Public Library of Science)
- Give away document readers, sell document writers (Adobe)
- Give women free admission, charge men (bars)
- Give children free admission, charge adults (museums)
- Give away listings, sell premium search (Match.com)
- Sell listings, give away search (Craigslist New York Housing)
- Give away travel services, get a cut of rental car and hotel reservations (Travelocity)
- Charge sellers to be stocked in a store, let people shop for free ("slotting fees" in supermarkets)
- Charge buyers to shop in a store, stock seller merchandise for free (membership stores such as Costco)
- Give away house listings, sell mortgages (Zillow)
- Give away content, sell information about the consumers (Practice Fusion)

- Give away content, make money by referring people to retailers (Amazon Associates)
- Give away content, sell stuff (Slashdot/ThinkGeek)
- Give away content, charge advertisers to be featured in it (product placement)
- Give away resume listings, charge for power search (LinkedIn)
- Give away content and data to consumers, charge companies to access it through an API (eBay)
- Give away "green" house plans, charge builders and contractors to be listed as green resources (FreeGreen.com)

FREE 3: FREEMIUM (SOME CUSTOMERS SUBSIDIZE THE OTHERS)

- Give away basic information, sell richer information in easier-to-use form (BoxOfficeMojo)
- Give away generic management advice, sell customized management advice (McKinsey and the McKinsey Journal)
- Give away federal tax software, sell state (TurboTax)
- Give away low-quality MP3s, sell high-quality box sets (Radiohead)
- Give away Web content, sell printed content (everything from magazines to books)
- Give away online games, charge a subscription to do more in the game (Club Penguin)
- Give away business directory listings, charge businesses to "claim" and enhance their own listings (Brownbook)
- Give away demo software, charge for the full version (most video games, which will allow you to play the first few levels to see if it's for you)
- Give away computer-to-computer calls, sell computer-to-phone calls (Skype)

- Give away free photo-sharing services, charge for additional storage space (Flickr)
- Give away basic software, sell more features (Apple QuickTime)
- Give away ad-supported service, sell the ability to remove the ads (Ning)
- Give away "snippets," sell books (publishers who use Google Book Search)
- Give away virtual tourism, sell virtual land (Second Life)
- Give away a music game, sell music tracks (Tap Tap Revolution)

ACKNOWLEDGMENTS

In the year and a half that I was writing this book, we had a baby (our fifth), I struggled with an unknown ailment that would eventually be diagnosed and then, a year later, undiagnosed as Lyme disease (which sucks as much as it sounds), flew more than 250,000 miles for speeches, continued to run *Wired*, and stupidly started yet another side-project company. That's a lot—too much. The fact that it was possible at all is entirely thanks to my wife, Anne, who not only served effectively as a single parent for much of the year, but did so with amazing grace and uncomplaining strength.

Anne not only shouldered the challenges of a big family and a traveling spouse, but when I was home she was the book's greatest champion. She was the one who would push me out the door on Saturday mornings to go write at the coffee shop, who read the pages late at night, and who got up with the baby in the morning so I could sleep in after late nights typing. The fact that this book felt both easy and fun to write is entirely due to her allowing it to be so by taking on so much, so generously. Of all the luck I have enjoyed in my life, nothing compares to having met Anne.

The other great debt of gratitude goes to my team at *Wired*, who brilliantly rose as I became an increasingly distant presence sending garbled sentence fragments over iPhones, and half-heard over speakerphones. The fact that we had another award-winning year is mostly due to Bob Cohn, Thomas Goetz, Scott Dadich, and Jake Young, who are the best team I've ever had the privilege to work with.

The book itself was also a collaboration, and I count myself incredibly fortunate to have not just one world-class editor but two. Will Schwalbe on the American side and Nigel Wilcockson on the British side didn't just redline the words. They brainstormed together in hour-long calls, and gave me the kind of sage advice and encouragement that only a true champion of a project can provide. They were both coaches and cheerleaders, if that does not mix sports metaphors too hideously. Anyway, they made this book much, much better, and there can be no greater praise for an editor than that.

I was also lucky to once again have Steven Leckart as my writing assistant. With my last book, we talked the chapters through, recorded them, and used the edited transcripts as raw material that I could use as a starting point. This time, perhaps because it was my second and I had a better idea of how to do it, or perhaps because the shape of the book was clearer in my head, we spent most of our time on fine-grained organization. Steven also researched and drafted most of the sidebars. I also had the help of Ben Schwartz, who appeared as a fifteen-year-old boy with Long Tail taste in my last book, and is now a college freshman with an appetite for science fiction. He read a heap of sci-fi books and digested their takes on abundance ("post-scarcity economics") for me with analytical wisdom far beyond his years.

Thanks again to Scott Dadich, *Wired*'s creative director, who designed both the cover of the paperback edition of *The Long Tail* and the bold magazine cover version of *Free*, and to Carl DeTorres, who designed the charts and sidebars with grace and style, just as he did with *The Long Tail*.

My agent, John Brockman, was exactly the tireless champion that you'd want an agent to be. The sales and publicity teams at Hyperion and Random House UK rose to the challenge of Free with innovative events, creative economic models, and endless enthusiasm, which is all the more impressive when you think of how so much of the publishing industry views Free with fear and suspicion. And my own publicity team at *Wired*, led by Alexandra Constantanople and Maya Draison, found countless ways to spread the word, from interviews to salons.

Intellectually, I am especially indebted to two people: Kevin Kelly, whose book *New Rules for the New Economy* laid the groundwork for much of my thinking about Free, and Mike Masnick of Techdirt, whose day-to-day exploration, reporting, and evangelism of Free both informed and inspired this book. George Gilder, who did early work on the economics of semiconductors and the deeper meaning of Moore's Law, remains a huge influence on my thinking. And Hal Varian, Google's chief economist, has taught me more, through his generous time and prescient writings, than any of my college professors.

Finally, my thanks to all the hundreds of people who have written to me and commented on my writings with examples of Free, their own stories of how they have used it, and thoughts on the economic models around it. They inspire me, keep me honest, and ultimately influenced every line in this book. This past decade has been a collective experiment in charting the future of a radical price, and it's the countless pioneers whose lessons I've attempted to capture here who deserve the ultimate thanks.

INDEX